Springer Theses

Recognizing Outstanding Ph.D. Research

Aims and Scope

The series "Springer Theses" brings together a selection of the very best Ph.D. theses from around the world and across the physical sciences. Nominated and endorsed by two recognized specialists, each published volume has been selected for its scientific excellence and the high impact of its contents for the pertinent field of research. For greater accessibility to non-specialists, the published versions include an extended introduction, as well as a foreword by the student's supervisor explaining the special relevance of the work for the field. As a whole, the series will provide a valuable resource both for newcomers to the research fields described, and for other scientists seeking detailed background information on special questions. Finally, it provides an accredited documentation of the valuable contributions made by today's younger generation of scientists.

Theses may be nominated for publication in this series by heads of department at internationally leading universities or institutes and should fulfill all of the following criteria

- They must be written in good English.
- The topic should fall within the confines of Chemistry, Physics, Earth Sciences, Engineering and related interdisciplinary fields such as Materials, Nanoscience, Chemical Engineering, Complex Systems and Biophysics.
- The work reported in the thesis must represent a significant scientific advance.
- If the thesis includes previously published material, permission to reproduce this must be gained from the respective copyright holder (a maximum 30% of the thesis should be a verbatim reproduction from the author's previous publications).
- They must have been examined and passed during the 12 months prior to nomination.
- Each thesis should include a foreword by the supervisor outlining the significance of its content.
- The theses should have a clearly defined structure including an introduction accessible to new PhD students and scientists not expert in the relevant field.

Indexed by zbMATH.

More information about this series at https://link.springer.com/bookseries/8790

Nicola Secciani

sEMG-based Control Strategy for a Hand Exoskeleton System

Doctoral Thesis accepted by University of Florence, Italy

Springer

Author
Dr. Nicola Secciani
Department of Industrial Engineering (DIEF)
University of Florence
Florence, Firenze, Italy

Supervisor
Prof. Benedetto Allotta
Department of Industrial Engineering (DIEF)
University of Florence
Florence, Firenze, Italy

ISSN 2190-5053 ISSN 2190-5061 (electronic)
Springer Theses
ISBN 978-3-030-90282-7 ISBN 978-3-030-90283-4 (eBook)
https://doi.org/10.1007/978-3-030-90283-4

This Springer imprint is published by the registered company Springer Nature Switzerland AG
The registered company address is: Gewerbestrasse 11, 6330 Cham, Switzerland

Tons of people have my gratitude for their contribution to my experience at the University of Florence during the last three years. Time has silently flown by as I grew up personally and academically. Knowing in advance that these few lines will never be enough, I will still try to show how much I am grateful for the opportunity, teaching, trust, and all the good received.

Here we go! First and foremost, I am indebted to my advisor, Professor Benedetto Allotta. It all started in Spring 2014 when he gave up one of his lunch breaks to explain Bode diagrams to a poor undergraduate student who was entering the world of Robotics and Automation for the first time. The interest and passion he inspired me in that context have been the driving force behind this journey. The possibility he gives me to continue to work in the field of Robotics for Medicine and Health care is what fuels my itinerary.

Additional thankfulness goes to my co-advisor, Doctor Alessandro Ridolfi, for he has been the one who welcomed me within his workgroup at the Mechatronics and Dynamic Modeling Laboratory of the Department of Industrial Engineering. His insightful comments and guidance through each of my steps have been crucial in defining the path of my research.

No one can do anything without backup from a great team! That is why a huge thanks goes to my lab mates, past, and present. Thanks for having accepted me in my exuberance, for the stimulating discussions, and for all the fun we have had in these last three years. A special thanks to Matteo Bianchi, with whom I shared the passion for this project.

Keyrole has been played by the exoskeleton pilot, Massimo Bianchini. From the initial idea to the last tests, he has always been present with monumental patience. With his feedback and insights, he has genuinely accompanied my research activities, allowing them a remarkable quantum leap. Very little would have been possible without his priceless cooperation.

Staggeringly important have been, finally, my family, my girlfriend, and my lifetime friends. Mom and dad are the ones who made it possible to start this journey; sister, grandma, cousins, aunts, and uncles are the ones who embellished it with their presence. I could not be more grateful to them, for they have always supported me with warm love and endless backing. Thanks to Martina, my loving, supportive, and patient girlfriend, for she has been becoming vital not only for my course of study but also for my life. Lastly, I would like to express my sincere gratitude to all of my dearest friends as growing up with them has shaped the person I am today.

Nicola Secciani

To Giuliano,
uncle and friend.

Supervisor's Foreword

Can Robotics really impact the lives of people with chronic disabilities? And, on the other side, can such people trust Robotics enough to rely on it for their daily activities? These are some of the questions the field of Robotics for Medicine and Healthcare seeks to answer. Lately, the branch of Wearable Robotics has been largely investigated by many research groups for these very purposes. The design of wearable devices, specifically hand exoskeletons, that can assist people with chronic motor disabilities during their daily lives emerges as a challenging and intriguing topic. Assisting the hand arises to be particularly hard for the small dimensions and the vast number of different movements a healthy limb can perform. Over the last few decades, a growing number of solutions have appeared in the scientific literature. Among the many features that set them apart and can make a difference, e.g., type of architecture or actuation system, the control strategy plays a key role. The more reactive, intuitive, reliable, effective the control is, the more likely the user would spend most of the day with the device.

This work tackles the development of a control strategy for an assistive hand exoskeleton that identifies a significant trade-off between responsiveness, intuitiveness, reliability and effectiveness. The primary contribution of this thesis is to pay unusual attention to the wearability and the real-time action of the proposed solution as assistive devices should not hinder user's movements and be ready when they are needed. The exploitation of the residual muscular activity in the form of surface electromyography signals represents a promising starting point to guarantee intuitiveness. A particular machine learning-based algorithm trained on the specific user's capabilities assures not only responsiveness, reliability, and effectiveness but also a high level of assistance customization.

An accurate study of the theoretical background, a precise comparison with the state of the art, and a detailed description of the implementation of the developed strategy guide the reader from the preliminary drafting of the requirements list to the actual clinical trials. Hence, this work can constitute a helpful design reference for researchers who want to further investigate this kind of applications.

Florence, Italy
August 2020

Benedetto Allotta

Parts of this thesis have been published in the following documents:

Journals

Secciani, N., Topini, A., Ridolfi, A., Meli, E., Allotta, B., *A Novel Point-in-Polygon-Based sEMG Classifier for Hand Exoskeleton Systems*, IEEE Transactions on Neural Systems and Rehabilitation Engineering, 28 (12), art. no. 9291412, pp. 3158–3166 (2020).

Secciani N., Bianchi M., Meli E., Volpe Y., Ridolfi A., *A novel application of a surface ElectroMyoGraphy-based control strategy for a hand exoskeleton system: A single-case study*, International Journal of Advanced Robotic Systems, Vol. 16, Issue 1 (2019).

Secciani, N., Bianchi, M., Ridolfi, A., Vannetti, F., Volpe, Y., Governi, L., Bianchini, M., Allotta, B., *Tailor-made hand exoskeletons at the University of Florence: From kinematics to mechatronic design*, Machines, 7 (2), art. no. 22 (2019).

Book Chapters

Secciani, N., Pagliai, M., Buonamici, F., Vannetti, F., Volpe, Y., Ridolfi, A. *A Novel Architecture for a Fully Wearable Assistive Hand Exoskeleton System*, Mechanisms and Machine Science, 91, pp. 120–127 (2021).

Allotta, B., Bianchi, M., Meli, E., Ridolfi, A., Secciani, N., *A portable tailor-made exoskeleton for hand disabilities*, Wearable Robotics: Systems and Applications, pp. 177–191 (2019).

Secciani N., Bianchi M., Ridolfi A., Vannetti F., Allotta B., *Assessment of a hand exoskeleton control strategy based on user's intentions classification starting from surface EMG signals*, Biosystems and Biorobotics, Vol. 22, pp. 440–444 (2019).

Secciani, N., Bianchi, M., Meschini, A., Ridolfi, A., Volpe, Y., Governi, L., Allotta, B., *Assistive hand exoskeletons: The prototypes evolution at the University of Florence*, Mechanisms and Machine Science, 68, pp. 307–315 (2018).

Contents

Acronyms

ABS	Acrylonitrile Butadiene Styrene
ADC	Analog-to-Digital Converter
ADLs	Activities of Daily Living
AI	Artificial Intelligence
ANN	Artificial Neural Network
AT	Assistive Technology
CAD	Computer Aided Design
DIEF	Department of Industrial Engineering
DIP	Distal InterPhalangeal
DNN	Deep Neural Network
DOF	Degree Of Freedoms
DT	Decision Trees
EB	Entropy-Based
FD	Frequency Domain
FIC	Family of International Classifications Network
FN	False Negatives
FP	False Positives
GUI	Graphical User Interface
HDsEMG	High Density Surface ElectroMyoGraphy
HES	Hand Exoskeleton System
ICA	Independent Components Analysis
ICF	International Classification of Functioning, Disability and Health
IDE	Integrated Development Environment
ISPO	International Society for Prosthetics and Orthotics
IZ	Innervation Zone
KNN	K-Nearest Neighbors
MCP	MetaCarpalPhalangeal
MDM Lab	Mechatronics and Dynamic Modeling Laboratory
ML	Machine Learning
MoCap	Motion Capture
MU	Motor Unit

MUAP	Motor Unit Action Potential
PCA	Principal Components Analysis
PCB	Printed Circuit Board
PIP	Proximal InterPhalangeal
RF	Random Forest
RI	Rehabilitation International
ROM	Range Of Motion
RP	Random Projection
sEMG	Surface ElectroMyoGraphy
SMA	Spinal Muscular Atrophy
SVM	Support Vector Machine
TD	Time Domain
TN	True Negatives
TP	True Positives
UNECE	United Nations Economic Commission for Europe
UNIFI	University of Florence
USB	Universal Serial Bus
WHO	World Health Organization

Chapter 1
Introduction

We stand on the brink of a technological revolution that will fundamentally alter the way we live, work, and relate to one another.

K. Schwab. The Fourth Industrial Revolution. Foreign Affairs, 12th Dec. 2015

Over the last years, the "Fourth Industrial Revolution", also known as "Industry 4.0"—characterized by a range of new technologies that fuse the physical, digital and biological worlds together—has been increasingly impacting all disciplines, economies, and industries. It has even been challenging the idea about what it means to be human. Driven by the tide of this change, robots have more and more permeated many aspects of human life to such an extent that even a delicate area as the Healthcare System has started consolidating the use of robotic devices as part of the so-called Assistive Technology (AT).

Conventionally born at the dawn of the nineteenth century—when an Italian printer, Pellegrino Turri, designed a rudimentary typewriter for a blind friend—the AT has evolved over the years to obtain a first formal definition within the *Technology-Related Assistance for Individuals with Disabilities Act* to later become law under the US President Ronald Reagan in August 1988. Ten years later, the *United States Assistive Technology Act* has revisited what was written in the previous document to identify as AT, also called adaptive technology, any "product, device, or equipment, whether acquired commercially, modified or customized, that is used to maintain, increase, or improve the functional capabilities of individuals with disabilities."

Nowadays, the AT is a broad family of products adequately described within the International Standard ISO 9999. Its definition has been revised once again to match the International Classification of Functioning, Disability and Health (ICF) terminology. As a part of the World Health Organization (WHO) - Family of International Classifications Network (FIC), the ISO 9999 defines as AT any "product

N. Secciani, *sEMG-based Control Strategy for a Hand Exoskeleton System*,
Springer Theses, https://doi.org/10.1007/978-3-030-90283-4_1

04	**ASSISTIVE PRODUCTS FOR MEASURING, SUPPORTING, TRAINING OR REPLACING BODY FUNCTIONS** Products that monitor or assess a person's medical condition, and products that support, or provide a substitute for, a specific body function Included are, e.g. products used in "medical treatment". Excluded are assistive products used exclusively by healthcare professionals. Assistive products for education and for training in skills, see 05 Assistive products attached to the body for supporting neuromusculoskeletal or movement related functions (orthoses) and replacing anatomical structures (prostheses), see 06 Tilting tables, see 05 36 06 Assistive products for seeing, see 22 03 Assistive products for hearing, see 22 06
04 03	**Assistive products for respiration** Equipment for assisting a person to breathe Vibrators, see 04 27 12 Assistive products for environmental improvement, see 27 03
04 03 03	*Inhaled-air pretreatment devices* Devices that take in air from the surrounding area and physiologically treat it before inhalation by warming, cooling or cleaning the air

Fig. 1.1 An excerpt from the ISO 9999 that shows how the classification is made: codes (on the left), titles, notes, and references (on the right)

(including devices, equipment, instruments, and software) especially produced or generally available, used by or for persons with disability to protect, support, train, measure or substitute for body functions, structures and activities, or to prevent impairments, activity limitations or participation restrictions". The same International Standard also offers a detailed three-level hierarchical classification—classes, subclasses, and divisions—of the AT products based on their function. Each class, subclass, or division consists of a three-pair digit code, a title, and notes and references to other parts of the classification if necessary (see Fig. 1.1).

Six are the ISO 9999 editions that have been produced from 1992 to 2016 thanks to the collaboration with the International Society for Prosthetics and Orthotics (ISPO), the Rehabilitation International (RI), the United Nations Economic Commission for Europe (UNECE), and the WHO. Each version aimed to revise wrong titles or notes and add new ones in step with the latest technological trends. The last edition, the ISO 9999:2016, has 945 titles, of which about 44 are new, and 456 have been revised. It is worth noting that the standard reports that "all of the assistive products in this classification are primarily intended for use outside of health care settings; however, some of the products can be used in facilities such as rehabilitation centers". Furthermore, there is also a specific reference to using such devices in settings where employment-related activities are usually performed.

Recently, within its last reports [1, 2], the WHO has pictured a detailed scenario of the current and future need of AT products all around the world. What emerges from these documents is a considerable demand for innovative solutions to tackle the

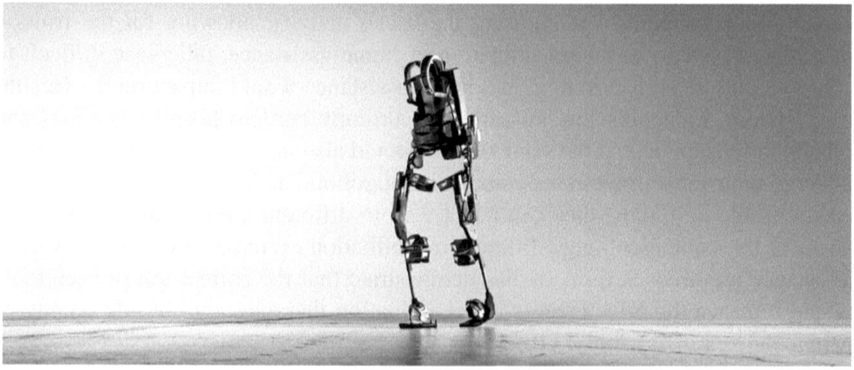

Fig. 1.2 An example of an assistive robotic device: a lower limb exoskeleton designed to assist people with motor disability during the gait

issue of disability. It is estimated that today more than one billion people need one or more assistive products—these numbers are even destined to double by 2030 due to the global population aging—however, only one out of ten people in need has access to assistive technology. The leading causes are high costs and a lack of awareness, availability, trained personnel, policy, and financing.

Such demographic, economic, social, technological, environmental, and political factors (DESTEP factors[1]) have paved the way to the advent of the Robotics for Medicine and Healthcare as a potent tool to overcome the availability limits of the standard AT [3, 4]. This branch of robotics is considered "the domain of systems able to perform coordinated mechatronic actions (force or movement exertions) on the basis of the processing of information acquired through sensor technology, with the aim to support the functioning of impaired individuals, rehabilitation of patients, care and medical intervention of patients and also to support individuals in prevention programs". It can be divided into five different classes: (i) Robotics for medical interventions; (ii) Robotics supporting professional care; (iii) Robotics assisted preventative therapies and diagnosis; (iv) Robotic assistive technology; (v) Robotics for rehabilitation treatment. Examples of such systems are powered exoskeletons [5, 6], like the one shown in Fig. 1.2, active prostheses [7, 8], or controlled robotic tools for surgery [9].

Robotics for Medicine and Healthcare products offer many advantages over non-robotic ones. Robotic devices generally incorporate an automatic control system to provide an intuitive and safe utilization to "perform coordinated mechatronic actions". Control systems are usually composed not only of hardware with computing power, such as processors, but also of sensors, remote communication channels, and data storage capacity. The combination of these elements makes robotic devices

[1] The DESTEP analysis is a model used for strategic decision-making over a process. It deals with external macroscopic factors. DESTEP stands for Demographic, Economic, Socio-cultural, Technological, Ecological, and Politics. The purpose of such analysis is to cover all these six topics looking for strengths, weaknesses and possible correlations between them.

capable of continuously monitoring the user's activity, allowing for the fruition of remote services, as telerehabilitation or home assistance, otherwise difficult to achieve. Both telerehabilitation and home assistance would impact on the Healthcare System, lightening the pressure that currently burdens hospital facilities and rehabilitation centers. At the same time, it would also increase the time patients can undergo their rehabilitation sessions. Besides, robotic devices are in general highly customizable and, thus, they can adapt well to different users or also to different needs of the same user—e.g., different rehabilitation exercises or different levels of assistance required. Several studies demonstrate that the correct use of such tools would improve the effectiveness of rehabilitation therapies and the life quality of people in need of assistance [10, 11].

However, Robotics for Medicine and Healthcare also presents several flaws. Since it is still a relatively new discipline, the level of awareness in the population, the policies that manage and regulate its use, and the amount of investment and financing in this sector are so far meager. Consequently, the lack of a wholly developed dedicated industrial fabric for the production of these devices and their high level of manufacturing complexity cause their cost to remain very high and, thus, out of the reach of most of the people in need. Moreover, since such products are mainly designed to share the environment and physically interact with people with disabilities, they have to meet strict requirements regarding wearability, safety, and comfort. Therefore, they also represent one of the most significant open points and one of the most challenging aspects in the human-robot interaction field.

1.1 Motivation

In light of the aforementioned reasons, the research activity presented within these pages stems from the desire to intervene in the delicate field of Robotics for Medicine and Healthcare, aiming not only to develop innovative solutions in the field of robotic devices but also to democratize their use and dissemination thanks to the adoption of guidelines of simplicity and low-cost throughout the whole design process. The purpose of the work summarized in this thesis was to integrate and complete an existing HES prototype (which will be briefly described in the following sections) designed for home assistance of people with hand disabilities by realizing an innovative control system. Hence, the dissertation will primarily focus on the design, development, and testing of a dedicated control strategy that, exploiting surface ElectroMyoGraphy (sEMG) signals, classifies the user's motor intentions and actuates the HES to reproduce them.

This activity was as stimulating as it was scientifically relevant. If, from the engineering point of view, assisting the hand represents a significant challenge because of its compound anatomy and the broad set of movements it can perform. From the clinic perspective, the hand is one of the most essential independence providers in everyday life. The loss of its functions is seen as one of the most impairing disabilities [12].

The work described here parallels the mechanical design of the HES performed by a colleague during his Ph.D. period. A thorough description of its mechanical characteristic is reported in [13]; the reader is encouraged to read it for more information on this topic. In this thesis, only some basics on the mechanics of the device will be given, while the development of the control system will be discussed in more detail.

1.2 Overall Framework

The work behind this thesis has been carried out at the Mechatronics and Dynamic Modeling Laboratory (MDM Lab) of the Department of Industrial Engineering (DIEF) of the University of Florence (UNIFI), which is operative in the field of Robotics for Medicine and Healthcare since 2013. The journey through the challenges and the open points of this subject began in that year with the development of the very first prototype of a Hand Exoskeleton System (HES). That exoskeleton was designed for a specific user: a person suffering from Spinal Muscular Atrophy (SMA) who proposed and voluntarily enrolled in the activity. Therefore, the whole design process has been centered on his particular needs and requirements. That preliminary version represented, nonetheless, the first embodiment of concept that has become the starting point for the improvements made in the following years.

Driven by the need to enlarge the target group of the developed technology and test it in a real-case scenario, a collaboration with the Don Carlo Gnocchi Foundation[2] was born in 2016—particularly with its rehabilitation center in Scandicci (Florence): the IRCCS Don Gnocchi. Since then, the Foundation has made the clinical expertise of its professionals, the use of equipment within the center, and the possibility to recruit patients for the joint research available to the MDM Lab researchers. The same collaboration, which become part of the current research of the rehabilitation center, has continued over the years until today, providing fertile ground for the development of two other versions of the HES prototype and the birth of this thesis work. Despite the exploitation of sEMG signals, as a non-invasive control technique for wearable devices, is a well-known procedure, literature is still lacking in studies carried out on actual patients outside the laboratory walls. At the time of writing, preliminary studies have been successfully concluded on an actual patient. Other subjects are being recruited for testing campaigns, which will be the object of future projects.

As for the MDM Lab current research activity, two are the ongoing projects which involve the use of the HES (Fig. 1.3): one, titled "HOLD: Hand exoskeleton system,

[2] The Don Gnocchi Foundation carries out rehabilitation activities and service for the elderly people in 28 residential structures and about thirty clinics spread in 9 Italian regions. https://www.dongnocchi.it.

for rehabilitation and activities Of daily Living, specifically Designed on the patient anatomy", funded by the University of Florence; the other, titled "BMIFOCUS: Brain Machine Interface in space manned missions: amplifying FOCUSed attention for error counterbalancing", funded by the Tuscany Region.

The two-year project HOLD aims to develop a HES that moves away from the laboratory to get as close as possible to the clinical environment in order to deal with real-case scenarios. Dysfunctions to the use of the hands can have pathological causes of a chronic (e.g., degenerative diseases) or incidental nature (e.g., strokes); in both cases, both rehabilitation and assistance are equally important since they are complementary in their purpose. While assistance aims to confer greater independence during the Activities of Daily Living (ADLs) and to increase the overall social interaction capabilities, on the other hand, the rehabilitative use of such a device focuses on improving the outcome of the therapeutic approach. This project was born with a very practical spirit. Thus, the expertise of the researchers of the DIEF is merged with the expertise of the professional of the IRCCS Don Gnocchi to center the design process on the actual needs and requirements of real patients suffering from hand disabilities. Started after the realization of the first exoskeleton prototype, the project focused in particular on improving its wearability, portability, and ease of use employing new mechanical design techniques—based on accurate ergonomic and structural studies to promote the prolonged use of the device—and through bio-inspired and flexible control strategies—to allow an intuitive and natural management of the device, both for rehabilitation and assistance. Finally, the fact that the resulting exoskeleton shall be much cheaper than other commercially available devices is another crucial point of the project.

The other two-year project BMIFOCUS aims to build up an integrated training platform to study the cognitive, sensory, and locomotor systems of pilots in extreme conditions such as piloting a spacecraft or a rover. Through augmented reality, the end-users will be trained to manage stress and emotions in the face of possible contingencies. Such a system will be composed of four subsystems: (i) a cerebral neurostimulation system which uses transcranial electrical stimulation techniques based on real-time analysis of biomedical signals to monitor and stimulate when needed the cognitive capabilities; (ii) an integrated acquisition system of electrophysiological and biomechanical signals; (iii) a simulation system of piloting tasks that interfaces a virtual reality sub-module with a haptic platform to reproduce one or more case scenarios through combined immersive experiences (motor, tactile, visual); (iv) a robotic rover to validate and test the whole system. In this framework, a bilateral HES shall be integrated with a training platform to reproduce the interaction forces with the virtual environments on the user's hand, e.g., during object handling.

Fig. 1.3 Logos of the ongoing projects at the MDM Lab

1.3 State of the Art

Nowadays, HESs are proven solutions capable of speeding up the recovery process of the hand by giving the possibility to carry out high-intensity rehabilitative sessions, otherwise not achievable through standard physiotherapy [14]. Such systems can also monitor and record the patients' signs of progress by measuring suitable anatomic parameters, allowing for a more accurate evaluation of the rehabilitation status. Unfortunately, not always is hand function recovered even after the complete treatment. There are cases [15–18] in which post-stroke patients cannot regain the original dexterity of their affected arm. In such situations, the hand exoskeletons can still play a critical role by providing partial or complete assistance to the impaired limb. For both uses, the stringent requirements on weight and dimensions—which limit the electronics, the transmissible power, and the use of standard mechanical components—have meant that the currently available HESs do not show an outcome as positive as expected.

Many are the possible classifications for hand exoskeletons based on their mechanical features, e.g., mechanical structure, number of Degree Of Freedoms (DOF), actuation system, and so on [19–22]; the reader is invited to find further details on this matter in [13]. To have a proper understanding of the work presented in this thesis, it is indeed sufficient to know that the hand exoskeleton involved in the research activity is:

- *rigid*—which means that the motion is transferred to the hand by means of one or more kinematic chains composed of rigid links;
- *single-DOF*—which means that the aforementioned kinematic chains have only one actuated degree of freedom;
- *single-phalanx*—that is each kinematic chain acts on the corresponding finger through a single contact point on a specific phalanx;
- *electrically actuated*—which means that the motion source is one or more electric motors;

besides, it also important to state that:

- such HES is composed of four kinematics chains (henceforth called finger mechanisms) acting on all the fingers except for the thumb;
- the finger mechanisms' geometry is optimized on the user's anatomy to follow the natural finger trajectories as accurately as possible;
- the actuated DOF is the one allowing directly for the flexion and extension of the finger;
- the base of the exoskeleton is rigidly fixed on the back of the hand while each kinematic chains interact with the corresponding finger on the middle phalanx;
- only one electric motor is employed in the system and, thus, the independent motion of the fingers is not achieved.

The reasoning behind these design choices and a thorough discussion on them are reported in [13]. Besides, although it has not been the subject of this thesis, Chapter 2 will cover the main points of the system's mechanical architecture and will give a more comprehensive overview of the device.

Regarding this research activity, an accurate study of the state of the art concerning possible control strategies has been based on the fact that the device was already mechanically optimized to follow the trajectories of the fingers reliably. Therefore the literature analysis has focused on the fascinating field of recognizing and classifying the user's intention—regarded as the will to move a limb in a certain way—rather than searching for complete control strategies for robotic devices.

The latest trends in the literature indicate the use of sEMG signals as one of the most promising ways to classify the user's motor intentions [23–26]. Different techniques have been investigated in recent times. Such methods are, more or less, based on the same procedure [27]: (i) collect time series of sEMG signals and split them into labeled time segments of an appropriate length; (ii) apply a filtering action to each segment and extract, if needed, time-domain or frequency-domain features (or a combination of them); (iii) use those features to train specific classifiers running pattern recognition algorithms. Assuming that the existence of a two-way relationship between EMG patterns and the corresponding gestures is reasonable, the classifier output is then translated into appropriate low-level commands for the HES actuation system to reproduce the classified intention.

While literature is well stocked with studies on sEMG classifiers [28–37], not many are the available research works centered on the implementation of such methods to a real-use exoskeleton This difference, which may seem small, is instead significant. The sEMG pattern modifications due to the use of an exoskeleton, or solely to its wearing, are indeed not negligible. Also, many refined techniques suggested by literature are not applicable in most cases: constraints on the complexity (e.g., number of DOF and number of sensors), wearability (e.g., setup for sensors and actuators, as well as the necessity to be as accessible as possible to a hand-impaired user), responsiveness (e.g., real-time requirements achievable on embedded electronics), and cost of the HES are also inherent limits to the classification strategy. As it will be observable in detail, sophisticated techniques give way to more elementary approaches for the application to a real device. At the time of writing and to the best of the author's knowledge, few studies strictly related to this thesis topic can be found in literature. The following selection will be analyzed based on three main characteristics: (i) sensors setup, in terms of number, positioning, and application on the user; (ii) classification method, in terms of classification criterion, classifiable gestures, and exploited features; (iii) actuation protocol, discriminating between single-trigger and continuous-effort. Other aspects, such as the sensors typology and the assistive or rehabilitative use, will not be considered since the signals' characterization is not the subject of this very study.

Ochoa et al. (2011)—The control strategy presented in [38] exploits sEMG signals collected from two forearm muscular bands: the *extensor digitorum*[3] and from the *flexor digitorum*.[4] The number of sensors is not specified, and an external and trained supervisor performs their application on the patient. The control system, designed to classify only the opening and closing gesture of the hand, runs on a custom-made printed circuit board. It real-time compares the amplitude of each signal to a corresponding voltage threshold and classifies the relative intention every time one of the two limits is overcome. The device is then moved accordingly, and the entire Range Of Motion (ROM) is covered in percentile steps of 10%.

Ho et al. (2011)—The method described in [39] uses the muscular signal generated by the *extensor digitorum* to trigger the opening action of the HES. The management of the closure is instead entrusted to the signals coming from the *abductor pollicis brevis*.[5] A trained person is supposed to place one sensor on each muscle. The classification is performed by comparing the amplitude of the signals with a threshold set at 20% of the user's maximum voluntary contraction of the corresponding muscle. Once the user's intention, namely, an opening or closing gesture, is classified, the actuation system is triggered to execute the entire movement at once. The whole control system runs on embedded hardware, fully wearable.

Leonardis et al. (2015)—Here in [40], an approach involving both hands is proposed: a bilateral HES for rehabilitation. It is meant to be used when the patient has only one impaired hand since the driving sEMG signals are collected from the non-paretic limb. The involved muscular bands are those three mentioned in the previous works and the sensor application still needs to be performed by means an external intervention. However, one crucial difference concerning the two previous works is represented by the fact that the signals collection does not take place on the

[3] The *extensor digitorum* is a muscular band of the posterior forearm, responsible for opening the four long fingers by acting principally on the metacarpophalangeal joints. Under continued action, it also extends the wrist, and then the elbow.

[4] The *flexor digitorum* muscular band, located in the anterior compartment of the forearm, can be split in two muscles: the flexor digitorum superficialis and the flexor digitorum profundus. It can be considered the antagonist of the extensor digitorum, and, therefore, it produces opposing joints torques and opposite movements.

[5] The abductor pollicis brevis is a superficial muscle located on the palm, right under the base of the thumb. As the name suggests, it is responsible for the abduction of the thumb, defined as the movement towards the palm.

arm influenced by the exoskeleton. Thus, the exoskeleton itself does not modify the sEMG patterns on which the classifier works. The authors use a light neural network implemented on a desktop PC as the classifier in this work. Still, no reference is given to the actuation method.

Meeker et al. (2017)—In [41], the authors present the application of a commercially available armband to the control strategy of a HES: the elastic armband includes eight sensors. Although it does not need to be precisely positioned by a trained person, it still requires an external intervention to be worn by the patient. A random forest algorithm, fed with the raw signals from the eight sensors, is in charge of the classification of the only two classifiable movements: opening and closing of all the fingers. The classifier output is not directly translated into a command signal, but it is first processed through a median filter and then compared to two thresholds. In the paper, nor the hardware nor the actuation method are mentioned.

1.4 Contribution and Thesis Structure

Starting from the scenario mentioned above, the research activities presented within this thesis dealt with providing the MDM Lab's low-cost assistive hand exoskeleton prototype with an intuitive and reliable control strategy, which guaranteed the portability, wearability, and inexpensiveness of the device. The HES control system consisted of a two-button trigger connected in an open loop to the device, at the beginning of this work. The necessity of using one hand to operate the exoskeleton placed on the other, and the complete lack of feedback on the user's hand state made the employment of the device unintuitive as well as unsafe.

From the analysis of the state of the art, the acquisition and classification into appropriate motor intentions of the user's sEMG patterns arose as a feasible but also preferable and widely-studied way to replace the trigger system. The main reason was the non-invasive and extraordinarily intuitive type of control that could be achieved: sEMG signals can be indeed collected directly from the skin and are derivatives of the natural means exploited by the brain to control the body movement. However, even if the literature presents many promising solutions about sEMG classifiers, only a small number of studies about implementing such decision-making algorithms on HES can be found. Focusing just on the few related works reported in the previous section, it can be seen that: (i) they all present control techniques which are based, when specified, on the classification of only three of the possible hand gestures—resting and complete opening or closing; (ii) the independent control of the single finger is taken into account by none of them; (iii) the works reported by Ochoa et al. and Ho et al. are the only ones that explicitly indicate that the whole control system, including the classifier, runs on embedded hardware—a necessary condition for the HES to be portable and fully wearable; (iv) the number of total sensors varies in a range from two to eight and, in each case, they need to be positioned by an external person; (v) only the work of Ochoa et al. shows a continuous-effort actuation method, which, compared to the single-trigger ones, offers more control over the ROM.

The main contribution of the work proposed in this thesis is to offer a solution that, by embodying a suitable trade-off among the strategies mentioned above, overcomes some of the limits of the current state of the art and is ready to be part of a proper clinical trial after the approval of an ethics committee. Such a novel sEMG-based approach aims to merge the strengths of the already known techniques and introduce a new classification criterion never applied before to wearable robotics. In particular:

- it involves the use of sEMG sensors placed adequately in specific positions to maximize the selectivity of the measurements on the specific muscles and, thus, to reduce possible disturbances deriving from the activation of other muscle groups;
- it provides the possibility to classify only the three elementary hand motor actions (opening, closing, and resting) to keep the complexity of the whole classification system low and to exploit a classifier computationally light enough to be run on embedded controller boards;
- it is based on a continuous-effort actuation method to offer complete control over the entire ROM, to allow for intermediate poses of the hand, and to result as natural as possible;
- it is specifically designed for the clinical sector and, thus, it can be quickly and straightforwardly tuned to different patients' needs by non-technical staff.
- it is integrated into a patient-friendly, low-cost, and fully wearable HES.

Chapter 2 gives a detailed description of the state of the art of the HES developed by the MDM Lab of the UNIFI DIEF at the beginning of this thesis work. Since this research activity paralleled the mechanical development of the device, this chapter also gives an overview of the mechanical modifications and improvements the exoskeleton has undergone until today.

Chapter 3, after a brief introduction of the physiology of the sEMG signals and the base principles of the classification theory, reports all the work done towards the implementation of the new control strategy: the changes made to the electronics of the system, the choice of the classification algorithm, and the development of two dedicated Graphical User Interfaces (GUIs).

Chapter 4 the actual implementation and the testing of the new control strategy. An overview of the modifications to the old HES electronics, which also involved an onerous mechanical redesign of the device, will be given. The chapter will finally focus on the experimental tests that have been performed in a pilot study and their results.

Chapter 5 concludes the thesis, providing a critical analysis of the achievements and opening up scenarios for possible future developments.

References

1. Organization WH (2011) World report on disability, Technical Report
2. Organization WH (2014) WHO global disability action plan 2014-2021: better health for all people with disability, Technical Report
3. Butter M, Rensma A, Boxsel JV, Kalisingh S, Schoone M, Leis M, Gelderblom GJ, Cremers G, Wilt MD, Kortekaas W et al (2008) Robotics for healthcare: Final report, Technical Report
4. Gelderblom GJ, De Wilt M, Cremers G, Rensma A (2009) Rehabilitation robotics in robotics for healthcare; a roadmap study for the european commission. In: 2009 IEEE international conference on rehabilitation robotics, pp 834–838
5. Gopura RARC, Kiguchi K, Bandara DSV (2011) A brief review on upper extremity robotic exoskeleton systems. In: 2011 6th international conference on industrial and information systems, pp 346–351
6. Lovrenovic Z, Doumit M (2016) Review and analysis of recent development of lower extremity exoskeletons for walking assist. In: 2016 IEEE EMBS international student conference (ISC), pp 1–4
7. Toledo C, Leija L, Munoz R, Vera A, Ramirez A (2009) Upper limb prostheses for amputations above elbow: a review. In: 2009 pan American health care exchanges, pp 104–108
8. Weiland JD, Humayun MS (2014) Retinal prosthesis. IEEE Trans Biomed Eng 61(5):1412–1424
9. Mylonas GP, Vitiello V, Cundy TP, Darzi A, Yang G (2014) Cyclops: a versatile robotic tool for bimanual single-access and natural-orifice endoscopic surgery. In: 2014 IEEE international conference on robotics and automation (ICRA), pp 2436–2442
10. Zampolini M, Todeschini E, Bernabeu Guitart M, Hermens H, Ilsbroukx S, Macellari V, Magni R, Rogante M, Scattareggia Marchese S, Vollenbroek M, Giacomozzi C (2008) Telerehabilitation: present and future. Annali dell'Istituto superiore di sanitá 44(2):125–134
11. Smarr CA, Fausset CB, Rogers WA (2011) Understanding the potential for robot assistance for older adults in the home environment. Georgia Institute of Technology, Technical Report
12. Anderson KD (2004) Targeting recovery: priorities of the spinal cord-injured population. J Neurotrauma 21(10):1371–1383
13. Bianchi M (2020) Development and testing of hand exoskeletons. Springer Nature
14. Rosenstein L, Ridgel A, Thota BSA, Alberts J (2008) Effects of combined robotic therapy and repetitive-task practice on upper-extremity function in a patient with chronic stroke. Am J Occup Therapy: Offic Publ Am Occup Therapy Assoc 1(62):28–35
15. Heller A, Wade D, Wood V, Sunderland A, Hewer R, Ward E (1987) Arm function after stroke: measurement and recovery over the first three months. J Neurol Neurosurg Psychiatry 131(50):714–719
16. Wade D, Langton-Hewer R, Wood V, Skilbeck C, Ismail H (1983) The hemiplegic arm after stroke: measurement and recovery. J Neurol Neurosurg Psychiatry 46(6):521–524
17. Sunderland A, Tinson D, Bradley L, Hewer R (1989) Arm function after stroke. an evaluation of grip strength as a measure of recovery and a prognostic indicator. J Neurol Neurosurg Psychiatry 52(11):1267–1272
18. Nakayama H, Jørgensen H, Raaschou H, Olsen T (1994) Recovery of upper extremity function in stroke patients: the copenhagen stroke study. Arch Phys Med Rehabil 75(4):394–398
19. Heo P, Gu GM, Lee S (2012) Current hand exoskeleton technologies for rehabilitation and assistive engineering. Int J Precis Eng Manuf 13(5):807–824
20. Troncossi M, Mozaffari-Foumashi M, Parenti-Castelli V (2016) An original classification of rehabilitation hand exoskeletons. J Robot Mech Eng Res 1(4):17–29
21. Bos RA, Haarman CJW, Stortelder T, Nizamis K, Herder JL, Stienen AHA, Plettenburg DH (2016) A structured overview of trends and technologies used in dynamic hand orthoses. J NeuroEng Rehabil 13(1):62
22. Sarac M, Solazzi M, Frisoli A (2019) Design requirements of generic hand exoskeletons and survey of hand exoskeletons for rehabilitation, assistive, or haptic use. IEEE Trans Hapt 12(4):400–413

23. Singh RM, Chatterji S, Kumar A (2012) Trends and challenges in emg based control scheme of exoskeleton robots-a review. Int J Sci Eng Res 3(8):933–40
24. Chowdhury R, Reaz M, Ali M, Bakar A, Chellappan K, Chang T (2013) Surface electromyography signal processing and classification techniques. Sensors 13(9):12 431–12 466
25. Nazmi N, Abdul Rahman M, Yamamoto S-I, Ahmad S, Zamzuri H, Mazlan S (2016) A review of classification techniques of emg signals during isotonic and isometric contractions. Sensors 16(8):1304
26. Meng Q, Meng Q, Yu H, Wei X (2017) A survey on semg control strategies of wearable hand exoskeleton for rehabilitation. In: 2017 2nd Asia-Pacific conference on intelligent robot systems (ACIRS), pp 165–169
27. Rechy-Ramirez EJ, Hu H (2015) Bio-signal based control in assistive robots: a survey. Digit Commun Netw 1(2):85–101
28. Hudgins B, Parker P, Scott RN (1993) A new strategy for multifunction myoelectric control. IEEE Trans Biomed Eng 40(1):82–94
29. Sapsanis C, Georgoulas G, Tzes A (2013) Emg based classification of basic hand movements based on time-frequency features. In: 21st mediterranean conference on control and automation, pp 716–722
30. Scheme E, Lock B, Hargrove L, Hill W, Kuruganti U, Englehart K (2013) Motion normalized proportional control for improved pattern recognition-based myoelectric control. IEEE Trans Neural Syst Rehabil Eng 22(1):149–157
31. Nadzri AABA, Ahmad SA, Marhaban MH, Jaafar H (2014) Characterization of surface electromyography using time domain features for determining hand motion and stages of contraction. Aust Phys Eng Sci Med 37(1):133–137
32. Veer K, Sharma T (2016) A novel feature extraction for robust emg pattern recognition. J Med Eng Technol 40(4):149–154
33. Atzori M, Cognolato M, Müller H (2016) Deep learning with convolutional neural networks applied to electromyography data: A resource for the classification of movements for prosthetic hands. Front Neurorobot 10:9
34. Jaquier N, Calinon S (2017) Gaussian mixture regression on symmetric positive definite matrices manifolds: Application to wrist motion estimation with semg. In: 2017 IEEE/RSJ international conference on intelligent robots and systems (IROS), pp 59–64
35. Ryser F, Bützer T, Held JP, Lambercy O, Gassert R (2017) Fully embedded myoelectric control for a wearable robotic hand orthosis. In: 2017 international conference on rehabilitation robotics (ICORR), pp 615–621
36. Nakamura G, Shibanoki T, Kurita Y, Honda Y, Masuda A, Mizobe F, Chin T, Tsuji T (2017) A virtual myoelectric prosthesis training system capable of providing instructions on hand operations. Int J Adv Robot Syst 14(5):172988141772845
37. Wan Y, Han Z, Zhong J, Chen G (2018) Pattern recognition and bionic manipulator driving by surface electromyography signals using convolutional neural network. Int J Adv Rob Syst 15(5):1729881418802138
38. Ochoa JM, Kamper DG, Listenberger M, Lee SW (2011) Use of an electromyographically driven hand orthosis for training after stroke. In: 2011 IEEE international conference on rehabilitation robotics, pp 1–5
39. Ho N, Tong K, Hu X, Fung K, Wei X, Rong W, Susanto E (2011) An emg-driven exoskeleton hand robotic training device on chronic stroke subjects: task training system for stroke rehabilitation. In: 2011 IEEE international conference on rehabilitation robotics, pp 1–5
40. Leonardis D, Barsotti M, Loconsole C, Solazzi M, Troncossi M, Mazzotti C, Castelli VP, Procopio C, Lamola G, Chisari C et al (2015) An emg-controlled robotic hand exoskeleton for bilateral rehabilitation. IEEE Trans Hapt 8(2):140–151
41. Meeker C, Park S, Bishop L, Stein J, Ciocarlie M (2018) Emg pattern classification to control a hand orthosis for functional grasp assistance after stroke. In: 2017 international conference on rehabilitation robotics (ICORR), pp 1203–1210

Chapter 2
Background: First-Stage Device

The work described in this thesis started in 2016 from the need to implement a safe and effective control system on an assistive hand exoskeleton prototype [1–3]. Starting from an in-depth kinematic study of the finger mechanisms, the main features of such HES will be discussed to clarify its functioning and highlight the aspects on which the control system requirements (reported in Sect. 1.4) have been identified.

Section 2.1 deals with the closed-form forward kinematic study, to mathematically characterize the finger mechanism. Such a mathematical delineation has been the basis of an optimization procedure, implemented to modify the mechanism geometry to fit different hand sizes.

Section 2.2 describes the optimization procedure, giving an overview of the idea behind and presenting in short the process itself. Details are left to the in-depth analysis presented in [1, 4].

Section 2.3 finally describes the HES that embodies the aforementioned kinematic analysis and optimization procedure. The end of the section emphasizes the main flaws found according to the results of several intermediate clinical tests performed on a single patient at the IRCCS Don Gnocchi[1] under the supervision of health professionals.

2.1 Kinematic Synthesis

The first HES prototype was composed of five elements: the housing for motors and electronics, and four independent planar mechanisms to transmit motion to the index, middle, ring, and little finger. This section is dedicated to the kinematics of

[1] The Don Carlo Gnocchi Foundation has agreed to the execution of the tests upon the presentation of an informed consent form signed by the involved subject.

N. Secciani, *sEMG-based Control Strategy for a Hand Exoskeleton System*,
Springer Theses, https://doi.org/10.1007/978-3-030-90283-4_2

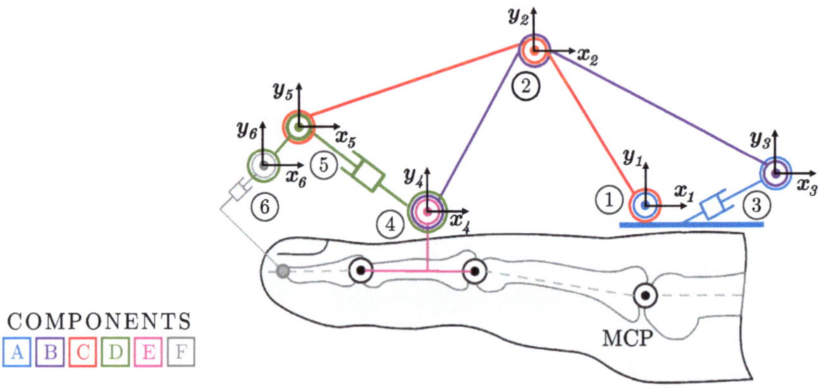

Fig. 2.1 The graphical representation of the 2D kinematic chain of the finger mechanism. Joints are identified by circled numbers, while the different links are named according to the colors legend in the low left corner

the mechanism (focusing on the index) and the scaling procedure implemented to adapt it to the other fingers. The possibility of using the scaling procedure to fit not only different finger dimensions but also different hand sizes will be finally discussed. This analysis's primary purposes were to mathematically characterize and numerically optimize a chosen single-DOF mechanism to make it reproduce the natural finger's trajectories as accurately as possible. The choice of optimize a single-DOF mechanism instead of opting for a multi-DOF one has been made according to the guidelines of wearability, simplicity, and cheapness imposed by the faced scenario. Such a decision has represented an effective solution in terms of functionality and manufacturability, leading to a smaller and lighter HES, easier to produce, actuate and control, and, thus, cheaper.

The study of the forward kinematics of the finger mechanism will be presented referring to what shown in Fig. 2.1, where:

- ①, ② and ④ are revolute joints;
- ③, ⑤ and ⑥ are pin slot joints;
- ① is the only actuated joint and controlled DOF;
- reference systems x_1y_1 is integral with component A, x_2y_2 and x_5y_5 with component C, x_3y_3 with component B, x_4y_4 with component D, x_6y_6 with component F and x_ey_e with component E;
- x_1y_1, whose correct positioning is at a known distance on the vertical above the MetaCarpalPhalangeal (MCP) joint of the hand, is the only fixed frame against which the pose of all the others is calculated;
- component A, E and F are respectively integral with the back of the hand, with the intermediate phalanx (henceforth also called middle phalanx), and with the distal phalanx;

- $\mathbf{O}_4$, namely the origin of x_4y_4 and actual end effector of the mechanism, is the main interaction point between the finger and the exoskeleton.[2]

The investigation will mainly focus on reconstructing the trajectory of the origin of frame x_4y_4 reported in frame x_1y_1 (${}^1\mathbf{O}_4$) as the joint 1 variable varies. Analytically knowing such a trajectory is indeed the first step to optimize the geometry of the whole finger mechanism, as it will be shown in the next section. However, since they will be required later, the calculation of the trajectory of the other remarkable points of the finger mechanism (${}^1\mathbf{O}_2$, ${}^1\mathbf{O}_3$, ${}^1\mathbf{O}_5$ and ${}^1\mathbf{O}_6$) will also be presented here below.

Analysing the kinematic chain of the finger mechanism the following eight independent relations can be identified:

$$\underline{{}^1\mathbf{O}_1} = {}^1\mathbf{O}_2 + \mathbf{R}_2^1\,\underline{{}^2\mathbf{O}_1} \tag{2.1}$$

$${}^1\mathbf{O}_2 = {}^1\mathbf{O}_3 + \mathbf{R}_3^1\,\underline{{}^3\mathbf{O}_2} \tag{2.2}$$

$${}^1\mathbf{O}_4 = {}^1\mathbf{O}_3 + \mathbf{R}_3^1\,\underline{{}^3\mathbf{O}_4} \tag{2.3}$$

$${}^1\mathbf{O}_5 = {}^1\mathbf{O}_2 + \mathbf{R}_2^1\,\underline{{}^2\mathbf{O}_5} \tag{2.4}$$

$${}^1\mathbf{O}_5 = {}^1\mathbf{O}_4 + \mathbf{R}_4^1\,{}^4\mathbf{O}_5 \tag{2.5}$$

$${}^1\mathbf{O}_6 = {}^1\mathbf{O}_4 + \mathbf{R}_4^1\,\underline{{}^4\mathbf{O}_6} \tag{2.6}$$

$$\underline{a_3}\,{}^1O_3^x + \underline{b_3}\,{}^1O_3^y + \underline{c_3} = 0 \tag{2.7}$$

$$\underline{a_5}\,{}^4O_5^x + \underline{b_5}\,{}^4O_5^y + \underline{c_5} = 0 \tag{2.8}$$

where the generic vector ${}^j\mathbf{O}_i = \left({}^jO_i^x\ {}^jO_i^y\ {}^jO_i^z\right)^T \in \mathbb{R}^3$ denotes the position of O_i in x_jy_j, the generic matrix $\mathbf{R}_i^j$ represents the rotation of x_iy_i with respect to x_jy_j, (a_3, b_3, c_3) and (a_5, b_5, c_5) are coefficients dependent on the linear constraints of the pin slot joints 3 and 5, and the terms referring to known geometrical quantities are underlined. Since the mechanism is planar, all the ${}^jO_i^z$ components are imposed equal to zero, and all the rotation matrices will result in elementary rotations about $\mathbf{z_i}$ axis through an angle α_i.

The kinematic state vector can now be defined as:

$$\mathbf{q} = \left[{}^1\mathbf{O}_2^T\ {}^1\mathbf{O}_3^T\ {}^1\mathbf{O}_4^T\ {}^1\mathbf{O}_5^T\ {}^1\mathbf{O}_6^T\ \alpha_2\ \alpha_3\ \alpha_4\right]^T \in \mathbb{R}^{18} \tag{2.9}$$

This vector contains the position of all the points of interest of the mechanism and three scalar angles of which one, α_2, represents the joint variable 1 and, thus, it is the

[2] Component F is an idle thimble which does not impact on the kinematic chain of the device. It is a passive element, added to accompany the motion of the distal phalanx.

control variable of the system. The elements inside the state vector can be calculated by solving the Eqs. 2.1–2.8 as function of $\alpha_2 \in \left[0\ \frac{\pi}{2}\right]$ itself and of all the geometrical quantities of the generic mechanism enclosed in:

$$\mathbf{S} = [{}^2\mathbf{O}_1^T\ {}^3\mathbf{O}_2^T\ {}^2\mathbf{O}_5^T\ {}^3\mathbf{O}_4^T\ {}^4\mathbf{O}_6^T\ a_3\ b_3\ c_3\ a_5\ b_5\ c_5]^T \in \mathbb{R}^{21} \tag{2.10}$$

In light of these premises, let $\tilde{\mathbf{q}} \in \mathbb{R}^{17}$ be the vector composed of all the unknown elements of $\mathbf{q}$ so that $\mathbf{q} = [\tilde{\mathbf{q}}\ \alpha_2]$, and let $\mathbf{f}(\alpha_2, \mathbf{S}) \in \mathbb{R}^{17}$ be the vector function composed by Eqs. 2.1–2.8, it is possible to solve the problem of the forward kinematics of the mechanism, $\tilde{\mathbf{q}} = f(\alpha_2, \mathbf{S})$, as follows. From henceforth, fixed geometrical quantities will no longer be underlined and every element of $\tilde{\mathbf{q}}$ vector (solved in terms of α_2, elements of $\mathbf{S}$ and already calculated terms) will have the corresponding equation boxed to indicate that it is a known quantity and, as such, shall be considered in the resolution of the subsequent calculations.

The vector Eq. 2.1 can be split into two scalar equations[3]:

$$0 = {}^1O_2^x + c\alpha_2{}^2O_1^x - s\alpha_2{}^2O_1^y \tag{2.11}$$

$$0 = {}^1O_2^y + c\alpha_2{}^2O_1^y + s\alpha_2{}^2O_1^x \tag{2.12}$$

and thus:

$$\boxed{{}^1O_2^x = -\left(c\alpha_2{}^2O_1^x - s\alpha_2{}^2O_1^y\right)} \tag{2.13}$$

$$\boxed{{}^1O_2^y = -\left(c\alpha_2{}^2O_1^y + s\alpha_2{}^2O_1^x\right)} \tag{2.14}$$

Then, dividing Eq. 2.2:

$${}^1O_2^x = {}^1O_3^x + c\alpha_3{}^3O_2^x - s\alpha_3{}^3O_2^y \tag{2.15}$$

$${}^1O_2^y = {}^1O_3^y + c\alpha_3{}^3O_2^y + s\alpha_3{}^3O_2^x \tag{2.16}$$

placed these two as a system with Eq. 2.7, the followings are obtained:

[3] As already mentioned, the z-component of all vectors is imposed equal to zero and, therefore, the calculation of such elements will not be taken into account.

$$ {}^{1}O_{3}^{y} = \frac{-A + \sqrt{(A)^2 - \left[\left(\frac{b_3}{a_3}\right)^2 + 1\right] H}}{\left(\frac{b_3}{a_3}\right)^2 + 1} \tag{2.17} $$

$$ {}^{1}O_{3}^{x} = -\frac{1}{a_3}\left(b_3\,{}^{1}O_{3}^{y} + c_1\right) \tag{2.18} $$

$$ s\alpha_3 = \frac{-{}^{1}O_{2}^{x} + {}^{1}O_{3}^{x} + \frac{{}^{3}O_{2}^{x}}{{}^{3}O_{2}^{y}}\,{}^{1}O_{2}^{y} - \frac{{}^{3}O_{2}^{x}}{{}^{3}O_{2}^{y}}\,{}^{1}O_{3}^{y}}{{}^{3}O_{2}^{y} + \frac{\left({}^{3}O_{2}^{x}\right)^2}{{}^{3}O_{2}^{y}}} \tag{2.19} $$

$$ c\alpha_3 = \frac{{}^{1}O_{2}^{y} - s\alpha_3\,{}^{3}O_{2}^{x} - {}^{1}O_{3}^{y}}{{}^{3}O_{2}^{y}} \tag{2.20} $$

where:

$$ A = \left(\frac{b_3}{a_3}\,{}^{1}O_{2}^{x} + \frac{b_3 c_3}{{a_3}^2} - {}^{1}O_{2}^{y}\right) \tag{2.21} $$

and:

$$ H = \left({}^{1}O_{2}^{x}\right)^2 + \left({}^{1}O_{2}^{y}\right)^2 - \left({}^{3}O_{2}^{x}\right)^2 - \left({}^{3}O_{2}^{y}\right)^2 + \frac{2b_3}{a_3}\,{}^{1}O_{2}^{x} + \left(\frac{c_3}{a_3}\right)^2 \tag{2.22} $$

Equations 2.3–2.5 can now be split into:

$$ {}^{1}O_{4}^{x} = {}^{1}O_{3}^{x} + c\alpha_3\,{}^{3}O_{4}^{x} - s\alpha_3\,{}^{3}O_{4}^{y} \tag{2.23} $$

$$ {}^{1}O_4^y = {}^{1}O_3^y + c\alpha_3 {}^{3}O_4^y + s\alpha_3 {}^{3}O_4^x \tag{2.24} $$

$$ {}^{1}O_5^x = {}^{1}O_2^x + c\alpha_2 {}^{2}O_5^x - s\alpha_2 {}^{2}O_5^y \tag{2.25} $$

$$ {}^{1}O_5^y = {}^{1}O_2^y + c\alpha_2 {}^{2}O_5^y + s\alpha_2 {}^{2}O_5^x \tag{2.26} $$

$$ {}^{1}O_5^x = {}^{1}O_4^x + c\alpha_4 {}^{4}O_5^x - s\alpha_4 {}^{4}O_5^y \tag{2.27} $$

$$ {}^{1}O_5^y = {}^{1}O_4^y + c\alpha_4 {}^{4}O_5^y + s\alpha_4 {}^{4}O_5^x \tag{2.28} $$

In order to proceed in the calculation of $c\alpha_4$ and $s\alpha_4$, the last two equations have to be placed as a system with Eq. 2.8 and, then, it can be solved similarly to the previous one, resulting in:

$$ {}^{4}O_5^y = \frac{-(b_5 c_5) + \sqrt{(b_5 c_5)^2 - \left(b_5{}^2 + a_5{}^2\right) T}}{b_5{}^2 + a_5{}^2} \tag{2.29} $$

$$ {}^{4}O_5^x = -\frac{1}{a_5}\left(b_5 {}^{4}O_5^y + c_5\right) \tag{2.30} $$

$$ c\alpha_4 = \frac{{}^{1}O_5^x + s\alpha_4 {}^{4}O_5^y - {}^{1}O_4^x}{{}^{4}O_5^x} \tag{2.31} $$

$$ s\alpha_4 = \frac{1}{\frac{\left({}^{4}O_5^y\right)^2}{{}^{4}O_5^x} - {}^{4}O_5^x}\left({}^{1}O_5^y - {}^{1}O_4^y - \frac{{}^{4}O_5^y}{{}^{4}O_5^x}{}^{1}O_5^x + \frac{{}^{4}O_5^y}{{}^{4}O_5^x}{}^{1}O_4^x\right) \tag{2.32} $$

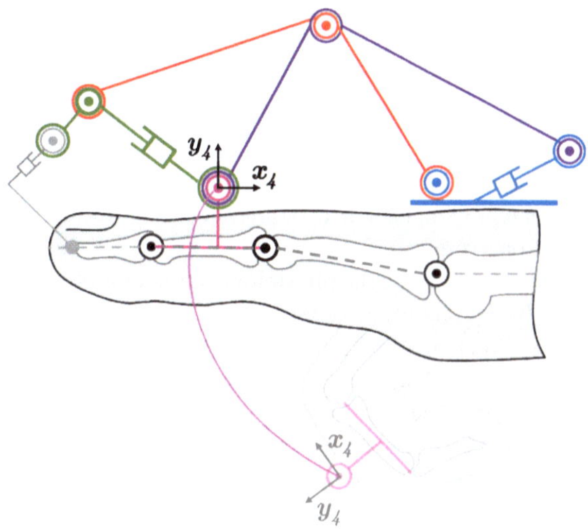

Fig. 2.2 In pink, the trajectory ${}^1\mathbf{O}_4(\alpha_2, \mathbf{S})$ with $\alpha_2 \in \left[0\ \frac{\pi}{2}\right]$

where:

$$T = -{a_5}^2\left[\left({}^1O_5^x - {}^1O_4^x\right)^2 + \left({}^1O_5^y - {}^1O_4^y\right)^2\right] + {c_5}^2 \tag{2.33}$$

Finally, the two scalar equations coming from the breakdown of the remaining Eq. 2.6 complete the kinematic analysis:

$$ {}^1O_6^x = {}^1O_4^x + c\alpha_4 {}^4O_6^x - s\alpha_4 {}^4O_6^y \tag{2.34}$$

$$ {}^1O_6^y = {}^1O_4^y + c\alpha_4 {}^4O_6^y + s\alpha_4 {}^4O_6^x \tag{2.35}$$

The kinematics of the mechanism is now completely described and the trajectory of ${}^1\mathbf{O}_4$ is reported in Fig. 2.2.

2.2 Optimization Procedure

Once identified the trajectory ${}^1\mathbf{O}_4(\alpha_2, \mathbf{S})$ through the analysis of the forward kinematics of the mechanism, the trajectory of the same point is calculated starting from the kinematics of the finger. By comparing these two, it is possible to study the coupling quality between the exoskeleton and the hand and, consequently, optimize the mechanism by minimizing the differences between the two trajectories.

As reported in Fig. 2.2, the connection point between the hand and the exoskeleton (identified as the point $\mathbf{O}_4$) can be assumed placed approximately on the middle-upper part of the intermediate phalanx. The first step of the optimization procedure is to track the natural motion of such a point through a Motion Capture (MoCap) analysis of the user's index finger. The exploitation of the MoCap technique to define the objective function to optimize is a remarkable advantage since, downstream of an entirely non-invasive procedure, the physiological finger motion can be tracked. Such tracking is possible by recording video images of markers placed on the lateral side of the right index finger in correspondence with the MCP joint, theProximal InterPhalangeal (PIP) joint, and the Distal InterPhalangeal (DIP) joint—the marker positioned on the MCP joint, assumed fixed in position during finger motion, constitutes the origin of the reference frame in which the trajectory of the other markers are calculated.

The acquired images are then processed through the open-source software Kinovea to extract the coordinate vectors of the trajectories of the markers as the finger moves. Henceforth, ${}^M\mathbf{O}_D$, ${}^M\mathbf{O}_P$, and ${}^M\mathbf{O}_4$ will indicate, respectively, the position of the DIP joint, the PIP joint, and the connection point calculated in $x_M y_M$. The position of the interaction point ${}^M\mathbf{O}_4$ cannot be directly extracted from the MoCap analysis but, assuming that component E and the intermediate phalanx constitute a single rigid body when the exoskeleton is coupled to the hand, it is possible to reconstruct the trajectory of a virtual marker placed on $\mathbf{O}_4$ starting from the trajectories of $\mathbf{O}_D$ and $\mathbf{O}_P$. In detail:

$$ {}^M\mathbf{O}_4^h = \frac{{}^M\mathbf{O}_D^h + {}^M\mathbf{O}_P^h}{2} + R_{z,\frac{\pi}{2}} \frac{m}{2} \frac{{}^M\mathbf{O}_D^h - {}^M\mathbf{O}_P^h}{||{}^M\mathbf{O}_D^h - {}^M\mathbf{O}_P^h||} \tag{2.36} $$

where the first addend represents the position of the midpoint of the second phalanx and the second addend is equivalent to a shift of half of the phalanx thickness (m) in the direction perpendicular ($R_{z,\frac{\pi}{2}}$) to the phalanx axis, namely the joining between DIP and PIP. The index h identifies a specific time instant because the trajectories of DIP and PIP joint are not analytical functions but time series of subsequent positions. However, in order to compare the trajectory just calculated with the one output by the forward kinematics other two more steps are required: (i) rewrite such trajectory in $x_1 y_1$; (ii) interpolate the new trajectory to get a continuous function to be then resampled at certain α_2 values. As shown in Fig. 2.3, the trajectory of ${}^M\mathbf{O}_4$ can be easily reported to $x_1 y_1$ by adding the displacement vector ${}^1\mathbf{O}_M$, which is a fixed and know quantity:

$$ {}^1\mathbf{O}_H^h = {}^M\mathbf{O}_4^h + {}^1\mathbf{O}_M \tag{2.37} $$

where $\mathbf{O}_H$ is the position of the interaction point imposed by the hand kinematics. Knowing such trajectory, it is possible to interpolate this point series as a function of α_2 by imposing the initial and the final point of the trajectory with $\alpha_2 = 0$ and $\alpha_2 = \frac{\pi}{2}$—respectively $(0, {}^1\mathbf{O}_H^0)$ and $(\frac{\pi}{2}, {}^1\mathbf{O}_H^f)$—and accordingly remapping all the

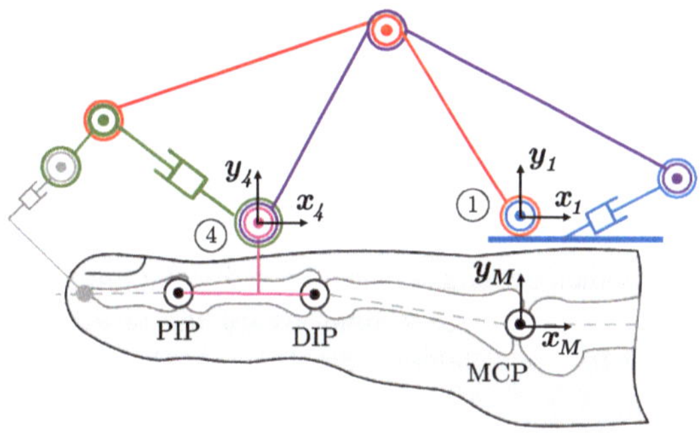

Fig. 2.3 The finger kinematic chain can be simplified into three revolute joints which, starting from the *metacarpus*, direct the motion of the three phalanges on the flexion/extension plane

other points within this interval. What it is obtained is an approximation of the trajectory ${}^1\mathbf{O}_H(\alpha_2)$.

Section 2.1 proves that, given $\mathbf{S}$ (Eq. 2.10), the vector that encloses the geometrical properties of the kinematic chain of the finger mechanism, it is possible to calculate in closed-form the trajectory of ${}^1\mathbf{O}_4$ (Eqs. 2.23)–(2.24) as a function of the joint variable α_2. The aim of the optimization procedure is instead to travel the other way around: the MoCap gives the trajectory of the contact point, ${}^1\mathbf{O}_H(\alpha_2)$, and the geometrical features in $\mathbf{S}$ become the free variables to be modified intending to make the finger mechanism track it as accurate as possible. Remembering ${}^1\mathbf{O}_1(\alpha_2)$ as the position of the contact point reproduced by the finger mechanism it is possible to define, firstly, the error between the two trajectories at a generic angular sample α_k (Eq. 2.38), and, secondly, an objective function weighting not only the average but also the maximum error (Eq. 2.39):

$$e_k = e\left(\alpha_k, \mathbf{S}\right) = \left| {}^1\mathbf{O}_H\left(\alpha_k\right) - {}^1\mathbf{O}_R\left(\alpha_k\right) \right| \tag{2.38}$$

$$f\left(\mathbf{S}\right) = \gamma \max_k |e_k| + (1-\gamma) \sum_{k=0}^{K} \frac{|e_k|}{K+1} \tag{2.39}$$

where γ, assumed as a scale coefficient to set the relative weight of the two errors has been set to 0.5 in this study to equally weigh maximum and average error. The optimization problem is hence defined as finding a particular vector of geometrical parameters $\mathbf{S}^{\star}$ which minimizes such objective function:

$$\mathbf{S}^{\star} = \min_{\mathbf{S}} f\left(\mathbf{S}\right) \tag{2.40}$$

Finally, because the nature of the problem itself was constrained—among the others, please remember the requirement of wearability of the HES—upper and lower boundaries have been imposed on each of the free variables. These constraints are chosen to rely on the user's anatomy and aim at defining a suitable geometry of the resulting mechanism which shall be safe for the user, practically manufacturable and not too bulky for the real use.

The constrained problem thus defined results to be nor convex nor linear because of the non-differentiability, non-continuity and the complexity of the objective function. The solution to such kind of problem is usually left to the recursive algorithm of numerical optimization. In this case, the applied techniques is the one proposed in [5] based on the Nelder-Mead simplex method [6]. The main idea behind this strategy is to perform a certain number of local searches for a sub-optimal solution, so that altogether, cover as much as possible the function domain. Each local search finishes when one or more conditions for the resulting simplex are met, and the starting point for the following search is probabilistically chosen to maximize the coverage of the domain. Once the number of iterations is over, the method examines all the convergence points to appoint the optimum.

2.3 The Original Prototype

What is reported in the previous section has paved the way to the actual manufacturing of a prototype of HES (Fig. 2.4) designed explicitly for a patient voluntarily enrolled in the study. Having suffered from SMA since birth, a neuromuscular disorder resulting in progressive muscle impairment, he had lost, among other body functions, the ability to extend hand fingers autonomously. Such a thing, as he claimed, was more limiting than other disabilities as it prevented him from comfortable object handling. This section will briefly discuss the details of the device developed to assist him.

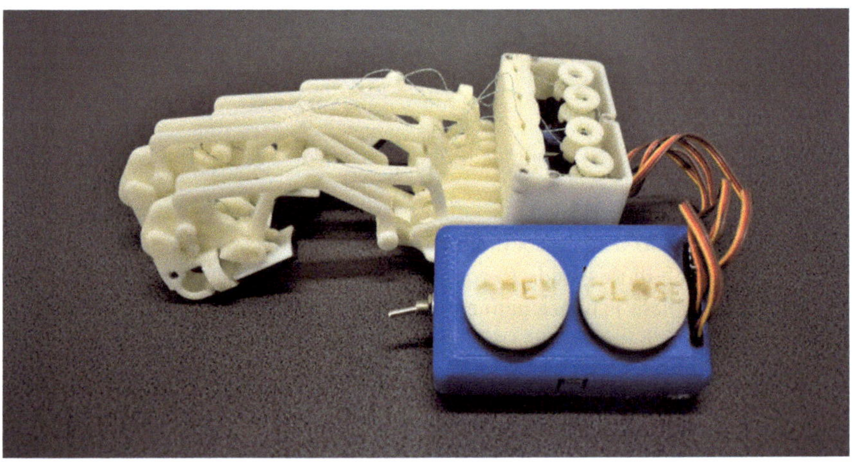

Fig. 2.4 The original prototype of the assistive hand exoskeleton system developed at the UNIFI DIEF. The figure, in addition to the mechanical structure and the actuation system, also shows the original control system (the blue box) composed of a switch, a controller board, and the two buttons that triggered the motion of the four motors

Before the beginning of the design process, a clinical assessment of the state of the patient's illness has been performed with the help of physiotherapists: the level of calcification of the finger joints was still low, and that the patient still had functional thumb mobility. Both of these points have proved to be of particular importance: the former, ascertained that the use of a finger-handling device would not have caused pain to the user, has given the green light to the mechanical design process; the latter has instead significantly lightened the process itself which could not have concerned, at least in the preliminary stages, a highly complex finger like the thumb.

According to the procedure presented in the previous section, the clinical assessment has been followed by the acquisition of the average two-dimensional trajectories of the user's index finger joints during the complete extension of the four long fingers. While the physiotherapist assisted the patient in performing several repetitions of the movement, video images of the patient's hand have been captured and processed using Kinovea. The acquired trajectories, extracted from the tracking of the markers placed on the index joints, have been then fed to the optimization algorithm leading to the definition of the first three-dimensional geometrical architecture of the index finger mechanism. Before moving on towards the manufacturing of such a mechanism, it has been virtually tested employing SOLIDWORKS® Motion. The mechanism has been paired with a 3D Computer Aided Design (CAD) model of the patient hand to qualitatively assess the tracking of the index finger's trajectory and the kinematic coupling and check for possible interpenetrations. Once finished the preliminary investigations, the geometry of the index finger has been scaled to match the dimensions of the other fingers. Then, the docking for the four finger mechanisms and the housing for the motors have been designed. Finally, just before the manufacturing phase, the components have undergone an additional reshaping: Fig. 2.4 shows that the shapes of the parts differ from the straight lines of the kinematic model. This final step, which has changed the aspect but not the kinematics, has been mandatory to avoid possible interpenetration with the user's finger. Moreover, integrating pins and shafts directly on the parts during this phase helped reduce each mechanism's lateral encumbrance remarkably. All the components have finally been 3D-printed in ABS exploiting a Dimension Elite printer by Stratasys.

As visible in Fig. 2.4, the mechanical structure comprises five parts: four independent finger mechanisms and a back frame designed as their docking station and to house the actuation system. The choice of 3D-printing all the device parts in ABS has allowed not to worry too much about the complex geometries that might have resulted from the optimization process—the additive manufacturing technology is indeed free from some of the manufacturing limits of the standard subtracting techniques. Besides, the choice of such a material has meant that the resulting device was already somehow inherently safe for the user: the ABS has mechanical characteristics that make it suitable for transmitting rigid motion for this kind of application but, at the same time, it gives all the parts an intrinsic mechanical clutch function against high loads that could harm the user. In other words, ABS components and joints would have broken before the loads transferred to the hand had become dangerous.

Following the requirements of lightness—indispensable for achieving a highly wearable device—high power density servo motors, one per finger mechanism, have been chosen as the core of the actuation system. The motion transmission from each actuator, housed in the fixed frame on the back of the hand, to the finger mechanisms, positioned over each finger, has been based on a cable that was wound or unwound by pulleys mounted on the motors shafts. It is worth noting that exoskeleton has been designed to intervene in the assistance of a hand unnaturally forced in a clenched fist deformity. Therefore, the exploitation of a cable transmission, capable of actively pulling the finger in extension and passively following them in flexion, fit perfectly into such a scenario. It also turned out to be one of the simplest and least cumbersome solutions. On the mechanisms themselves, holes have been reproduced for the passage and anchoring of the cables. The pulleys have been 3D-printed of different sizes, aiming to confer the different fingers with the same MCP joint angular velocity: the length of cable involved varies depending on the size of the finger. Moreover, since the length of cable necessary to achieve a complete motion of the finger resulted in being longer than the one unwound during a full rotation of the pulleys, these motors have been modified to allow for the continuous spinning. Such an adjustment was preferred to the design of sufficiently large pulleys as it would have compromised the compactness and the wearability of the system.

An Arduino Nano board, housed in a separate box placed on the forearm, controlled the actuators' behavior according to the triggers coming from two buttons (the two white circles in Fig. 2.4). A standard 6-Volt lithium battery pack—placed next to the shoulder using a Velcro band—powered the Arduino and the motors. The control strategy was simple: a very light code recursively checked for the state of the two buttons; if one of them happened to be pressed, the control board reacted by sending the servo command to the actuators that simultaneously started spinning at a fixed speed. The spinning direction depended on the specific button. This prototype did not include sensors to automatically control the finger mechanisms' position, and their ROM was manually managed by the user.

The whole mechatronic system presented above has been finally assembled and worn on the patient's hand using a sports glove—to which it had previously been sewed—as a base. The user has been told to perform several complete opening and closing gestures while video images were captured to be later processed with Kinovea. The markers have not been placed on the index finger joints, not easily accessible once the HES was worn, but instead stuck on the mechanism joints. The idea was to compare the acquired trajectories with the ones coming from the optimized kinematic model as a preliminary assessment of the accuracy of the device in reproducing the correct motion. Even though this qualitative comparison has given promising results, the tests conducted on this first version highlighted several flaws—some of those pointed out by the patient himself—which negatively impacted its practical usability: (i) the exoskeleton, not allowing for the abduction-adduction movement of the fingers, forced the user to move the fingers only on the flexion-extension plane resulting in a highly annoying feeling; (ii) having to use one hand to control the motion of the other because of the buttons-triggered control strategy resulted in being, as it could be expected, very unintuitive, unnatural and doubly clumsy; (iii) the absence of any type of position feedback over the flexion-extension angle of the MCP joint, and thus the lack of an automatic control over it did not make the user feel completely safe, causing additional mental stress and fatigue due to the necessary high amount of attention he had to pay continually.

In conclusion, the first-stage device ended up being compact (the HES did not exceed the hand size), fully wearable, light (it weighed just under 500 g), and cheap (it cost just over 500€); nonetheless, the user's feedback pointed out that it resulted still far from the actual clinical application and the practical use. Starting from such a review, this thesis work and other parallel activities—aiming at revamping the mechanical structure and, in particular, the actuation system—have begun; their results will be presented and discussed in the following chapters. The heart of this work will focus on the steps that have led to the development and implementation of an automatic control logic that overcame the limits arisen from testing this version of the HES.

References

1. Bianchi M (2020) Development and testing of hand exoskeletons. Springer Nature
2. Conti R, Meli E, Ridolfi A (2016) A novel kinematic architecture for portable hand exoskeletons. Mechatronics 35:192–207 April
3. Conti R, Meli E, Ridolfi A, Bianchi M, Governi L, Volpe Y, Allotta B (2017) Kinematic synthesis and testing of a new portable hand exoskeleton. Meccanica 52(11–12):2873–2897 January
4. Bianchi M, Fanelli F, Meli E, Ridolfi A, Vannetti F, Bianchini M, Allotta B (2018) Optimization-based scaling procedure for the design of fully portable hand exoskeletons. Meccanica 53(11–12):3157–3175 May
5. Luersen M, Le Riche R, Guyon F (2004) A constrained, globalized and bounded nelder-mead method for engineering optimization. Struct Multidiscip Optim 27(1):43–54 May
6. Nelder J, Mead R (1965) A simplex method for function minimization. Comput J 7(4):308–313 January

Chapter 3
The New Control System

Beyond the just reported control issues, an important thing that has also emerged during the tests was the accurate tracking of the acquired fingers' trajectories. This point was crucial for the choice of the new control strategy to implement. The availability of a device that already benefits from such mechanical optimization has indeed directed this thesis work towards the research for a suitable technique that would allow a reliable classification of the user's intentions to reproduce, leaving the motion management of the exoskeleton to a standard position and speed control.

When it comes to classifying the motor intentions of a person to control wearable robotic devices accordingly, the technical literature (Sect. 1.3) is evident in indicating the use of biological signals as the solution to be adopted for a control experience as intuitive and natural as possible. In particular, in the last years, the use of sEMG signals, i.e., epidermal measurement of the muscular activity in the form of electric potential, has been attracting the attention of more and more researchers. Surface electromyography is, in fact, an emerging discipline that boasts among its strengths the total non-invasiveness and, thanks to the technological progress of sensors and signal processing techniques, an excellent characterization of the muscular state.

This chapter will deal with the topics mentioned above, giving the basis to understand the path taken, discussing the choices made in detail, and reporting every step of the work done. While going through the chapter, it is essential to consider that the work phases described sequentially are, instead, the result of a circular process. Each step has hence been carried out in parallel and has influenced the other ones.

Section 3.1 describes the fundamental principles of the physiology of sEMG signals and the theory of classification. They both will constitute the theoretical background for the choices reported in the following sections.

Section 3.2 presents the electronics changes, which have been the basis for implementing the new control strategy. Each component is described in its functioning, and its main characteristics are stressed.

N. Secciani, *sEMG-based Control Strategy for a Hand Exoskeleton System*, Springer Theses, https://doi.org/10.1007/978-3-030-90283-4_3

Section 3.3 tackles the main focus of this work: the choice and the development of the new control strategy. Particular emphasis will be given to the original requirements, the various issues that have come up during the early stages of testing, and the efforts taken for their resolution. Ample space is also left to the description of a custom GUI designed to ease the process of sEMG signals collection and classifier setup.

3.1 Theoretical Background

This section aims to discuss the central theoretical notions on which the research work described in this thesis has been based. In particular, the principles of electromyography and classification will be addressed. Both are vast, complex, and above all, still in discovery subjects. What is written is not intended to be an exhaustive treatment of such topics but rather a summary of the information available in the literature, with particular attention to those areas that will be referred to later along the thesis.

3.1.1 sEMG Signal Physiology

Electromyography is the discipline that, from the beginning of the 20th century, studies and interprets the bioelectric signals coming from the neuromuscular system, also called myoelectric activity [1, 2]. The neuromuscular system is the apparatus humans use to interact with the environment; it comprises nerve endings in the corticospinal tract, (α)-motoneurons in the spinal cords and skeletal muscles. The set of an (α)-motoneuron and the innervated muscle fibers is called Motor Unit (MU); it represents the elementary units responsible for the contraction of muscles [3, 4]. When a motoneuron is excited by a nervous stimulus, the action potential propagation activates all muscular fibers supplied by its axonal branches; the propagation starts from the Innervation Zone (IZ) and moves in the opposite direction toward the extremities of the muscle. The number of MUs per muscle in humans varies roughly in a range from 100 to 1000; besides, their fibers are randomly distributed in the muscle and intertwined with fibers belonging to different MUs. For these reasons, their activation is observable only as an overlapping of the single contributes: this resulting signal goes under the name of Motor Unit Action Potential (MUAP). Collecting such signals is possible by placing in the proximity of the muscle one or more electrodes, which record the electrical potential generated by the electromagnetic field due to polarization and depolarization of muscle fibers.

In practice, there is much difference between the various electromyographic sensors commercially available. It is precisely their typology and their functioning that discriminates the nature of the signal that can be measured. Intramuscular sensors allow for the collection of classic EMG signals. Although this practice is decidedly invasive as needle-like electrodes are implanted directly into the muscle, the sig-

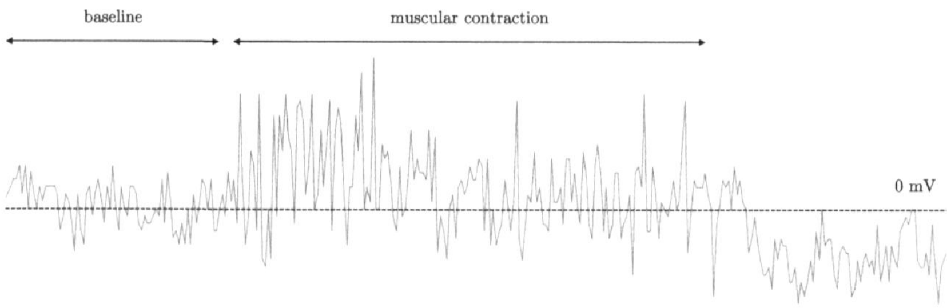

Fig. 3.1 An example of sEMG signal captured using Ag/AgCl electrodes

nals that are obtained are very selective and not very noisy. Epidermal sensors are instead designed to capture the electromyographic activity directly through the skin. This signal, which takes the name of sEMG, is characterized by a significantly higher noise level compared to the intramuscular one because it is filtered by layers of different tissue (e.g., fat, skin, other muscles) before reaching the electrode (see Fig. 3.1). Nevertheless, surface electromyography is in active growth since, being a completely non-invasive technique, it lends itself very well to scientific experimentation. The third category of electromyography is the High Density Surface ElectroMyoGraphy (HDsEMG), which uses sensors made of dense matrices of individual electrodes. The use of such sensors allows for a better spatial representation of the muscular activity, paving the way to the EMG decomposition—i.e., decompose the signals into the single MUAPs to observe the behavior of the motoneurons.

Surface electromyography, which is the technique exploited in this research work, will now be discussed in detail, presenting the main characteristics of a typical acquisition setup and the conditioning issues of sEMG signals.

The most used sEMG electrodes are made of a silver wire coated with a layer of solid silver chloride (Ag/AgCl). Unlike the other standard types of electrodes (silver chloride, silver, and gold), Ag/AgCl ones present a lower sensibility to the rubbing on the skin and, thus, usually lead to a cleaner signal collection, especially when soaked with an electrolyte gel. Properties of such contacts are an indeed higher signal-to-noise ratio and decent filtering of motion artifacts and power line disturbances.

Electrodes' dimensions vary from a few millimeters to some centimeters, and they can be of circular or rectangular shape. While the shape does not influence the acquired signal much, dimensions do matter: the larger the electrode, the lower the selectivity of the measurement.

EMG signals can be, in general, acquired in monopolar or bipolar disposition. Monopolar acquisitions reflect the electrical status of the skin right below the electrode. However, all the possible environmental (e.g., electromagnetic interference) and physiological (e.g., muscle crosstalk) disturbances also contribute. On the other hand, bipolar signals are acquired as the analogical difference between two monopolar signals. Using the difference of two signals, particularly when acquired from two sources very close to each other, means that the resulting measurement is less affected by the so-called common-mode disturbances, i.e., those that act in the same way on both monopolar signals. In light of the reasons mentioned above, not only the EMG collection is usually performed in the bipolar configuration, but in general, an additional electrode, common to all bipolar couples, is also added in a skin area free from muscular activity to serve as a shared reference of the overall electrical level of the body.

It is observable that, due to the bipolar acquisition methods and to a susceptible skin interface, electrode positioning and skin preparation are crucial for a good quality sEMG collection. For example, placing a bipolar sensor symmetrically with respect to the IZ and on the propagation direction of the action potential will result in a null signal detection since the two electrodes will record the same action potential but with the opposite sign. Recommendations for the correct positioning [5] suggest placing the sensor in the area between the IZ and the tendon region to avoid such problems. Regarding the improvement of the electrode-skin contact, the removal of body hair and layers of dead epidermal cells, and the sprinkling with electrolyte liquid of the affected skin areas are indicated as the steps to the correct skin preparation.

Whatever the type of surface electrodes used, the myoelectric signals appear on the skin in the form of an electric potential difference of limited amplitude (in the range of microvolts and millivolts) and bandwidth (10 500 Hz). Such signals also have a stochastic behavior, present a low signal-to-noise ratio, and are affected by several disturbances (e.g., interference from the electric power line, electromagnetic pollution, motion, and electrocardiographic artifacts and crosstalk). For these reasons, they are particularly tricky to collect, and very accurate equipment is required for its measurements. Commonly-used Analog-to-Digital Converters (ADCs) have an operational range from a few up to 10 V; this means that sEMG signals have to be adequately amplified before their digital conversion. After the sampling, which, according to the Nyquist-Shannon theorem, should occur at a frequency equal to or higher 1000 Hz, the signals are processed by a differential amplifier, which is also responsible for rejecting the common-mode disturbances. Other undesired phenomena, as the aliasing, motion artifacts, and parasitic frequencies, can be attenuated after the digitization using appropriate digital filters [6, 7].

However, the information contained in the raw sEMG signals remains challenging to interpret even after all the mentioned processing. Hence, it is a common practice to split the signals into epochs (or time windows), lasting from 250 to 500 ms [8], and extract representative indices (or features) from them. These indices can be calculated in the Time Domain (TD) directly from raw sEMG signals, or in the Frequency Domain (FD) derived from signals' spectrum analysis. TD features extraction is usually easier to implement, not requiring prior signal processing. Nevertheless, sEMG signals are strongly non-stationary, and this introduces errors and uncertainty in the calculation of the TD features since it assumes the data to be stationary. Finally, in the few particular cases in which neither the TD nor the FD features prove representative for the problem in question, it is possible to use the so-called Entropy-Based (EB) features based on the concept of entropy of a stochastic signal.

Since the last type is only used in very uncommon cases, EB features will not be further explored in this thesis.

Well known examples of TD features are the Mean Absolute Value (MAV), also known as Average Rectified Value (AVR), and the Root Mean Square (RMS) defined as:

$$MAV_e = \frac{1}{N_e}\sum_{k=1}^{N_e}|s_k| \qquad RMS_e = \sqrt{\frac{1}{N_e}\sum_{k=1}^{N_e}(s_k - MAV_e)^2} \tag{3.1}$$

where the subscript e highlights the quantities that depend on the length of the epoch considered, N_e is the number of samples within the considered epoch, and s_k is the k-th signal sample of the epoch. Both these indices are very similar, and both are used to express the amplitude of the signal, but their physical meaning is different: the MAV can be interpreted as the myoelectric activity level; the RMS gives information about the power of the sEMG signal itself.

In the FD, common features are instead the Mean Frequency (MNF) and Median Frequency (MDF), which give information on the power distribution of the signal. Both are based on the assumption of having already calculated the power spectral density function, $P(f)$, that describes the power associated with the different signal frequencies. MNF and MDF are defined as follows:

$$MNF_e = \frac{\sum_{f=0}^{f_s/2} f P_e(f)}{\sum_{f=0}^{f_s/2} P_e(f)} \qquad MDF_e \ s.t. \sum_{f=0}^{MDF_e} P_e(f) = 0.5 \tag{3.2}$$

where f is the generic frequency and f_s is the sampling frequency.

The four listed above are just some typical examples of the many indices that can be extracted from the electromyographic signal to study its information content. An overview on other possible sEMG features, according to the most recent data available in literature [9–11], is given in Table 3.1.

3.1.2 *Classification Theory*

Classification theory is a section of the model theory comprising all the doctrines regulating the grouping of objects based on their similarities, differences, or according to a given set of principles. Such a set of principles usually depends upon the nature of the objects themselves. The basic rules for organizing a domain of objects into classes require that such division should not leave two classes with any object in common. At the same time, all of the objects of the domain should be contained in one class. However, there are cases in which some objects can, with equal correctness, be accepted into two different classes. Classification theory, as it has just been described, can be applied in all the branches of knowledge, but, as it can be expected, it assumes different characteristics and rules depending on the context. In the fields of

Table 3.1 Commonly-used features in surface electromyography [9–11]

Feature name	Acronym	Methodology
Integrated EMG	IEMG	TD
Mean Absolute Value	MAV	TD
Modified Mean Absolute Value (Type 1)	MMAV	TD
Modified Mean Absolute Value (Type 2)	MMAV2	TD
Mean Absolute Value Slope	MAVSLP	TD
Simple Square Integral	SSI	TD
Variance	VAR	TD
Root Mean Square	RMS	TD
Waveform Length	WL	TD
Zero Crossing	ZC	TD
Slope Sign Change	SSC	TD
Willison Amplitude	WAMP	TD
Autoregressive Coefficients	AR	TD
Mean Frequency	MNF	FD
Median Frequency	MDF	FD
Modified Mean Frequency	MMNF	FD
Modified Median Frequency	MMDF	FD
Shannon entropy	H_{sh}	EB
Fisher entropy	H_{fi}	EB

Computer Science and Robotics the term classification is strictly related to Artificial Intelligence (AI) and Machine Learning (ML) [12]. In particular, Classification is an actual division of supervised learning, which in turn is a subcategory of the ML together with the unsupervised learning. Nonetheless, classification problems can be found and addressed in each of the ML branches. Classifiers aim at labeling new instances in specific categories based on training performed on past observations.

Aconventional classification problem is tackled in three steps: (i) data collection and preprocessing; (ii) model training and validation; (iii) model testing. The first step plays a vital role since the quality, and the phenomenon representativeness of the acquired dataset are crucial to obtain acceptable and consistent solutions to the problem. Moreover, data is rarely presented in the proper format to be correctly processed by the model. There can be cases in which the classifier algorithm needs some features to be extracted from the raw data to improve its performance, others in which the required features are not independent; the meaningful information has thus to be separated from the redundant one. In each of the cases mentioned above, preprocessing is a mandatory activity. It can take the form of normalization, filtering action, features selection, and dimensionality reduction technique, e.g., Principal Components Analysis (PCA), Random Projection (RP), or Independent Components

Analysis (ICA), or of a labeling procedure. This first phase ends by dividing the original dataset into three subsets: a training and a validation set, used in the second phase and a test set, which is the protagonist in the third phase.

Before continuing with the explanation of the second and third stages, it is essential to remember that different classifiers are distinguished not only among the various classes but also within the same class. Each classifier has its parameters and hyperparameters indeed. In general, it is possible to affirm that, once fixed the hyperparameters—which define the morphology of the classifier—the parameters—which strongly depend on the hyperparameter selection—establish the classification performance.

Firstly, in the model training and validation phase, a metric for measuring the model performance is selected. However, it is usually recommended to have more than one metric: e.g., accuracy (i.e., the percentage of the correctly classified instances over the total number of them) is commonly used, but also the time required to perform a single classification is another critical value for real-time applications that might have to be taken into account. Then, different algorithm morphologies, obtained by exploiting distinct parameter and hyperparameter configurations, are trained over the training set until an acceptable level of the chosen metric(s) is reached and the best—the one with the highest value—is preferred. Finally, the parameters of the trained models are fine-tuned to fit the validation dataset, which, being different from the training one, allows the learning process state to be monitored to avoid the "overfitting" issue. Overfitting is defined as a sort of excessive specialization of the classifier on the training data, which makes the classifier exceptionally performing only on the data with which it has trained. The training and validation strategy, i.e., how parameters and hyperparameters are recursively tuned for improving the model performance, is heavily dependent on the class of the model itself.

During the last phase, the focus shift to the generalization capability of the best model identified by the previous steps. Generalization stands for the ability of the classifier to adapt to unseen data; as a consequence, it also indicates the ability to generalize the prediction regardless of the type of input data. Qualitatively speaking, it is the opposite of overfitting and results as a key feature in providing an idea of the classifier performance within an actual application task. This third step is carried out by feeding the classifier with the test set, the one subset which has been kept aside since right after the first passage. Typically, the classifier's generalization performance is assessed using the same metric(s) of the training and validation phase.

When it comes to ML classification metrics, it is noteworthy to introduce first the concept of Confusion Matrix. Confusion matrices are useful tools to evaluate the performance of ML classification models, represented by tables with different combinations of predicted, in columns, and actual, in rows, values (see Fig. 3.2). As a matter of fact, from the intersection of rows and columns, it does arise the possibility to compute the desired metrics. The interpretation of the labels reported in the table is the following:

- True Positives (TP): the number of times the model predicts the condition when the condition is present;

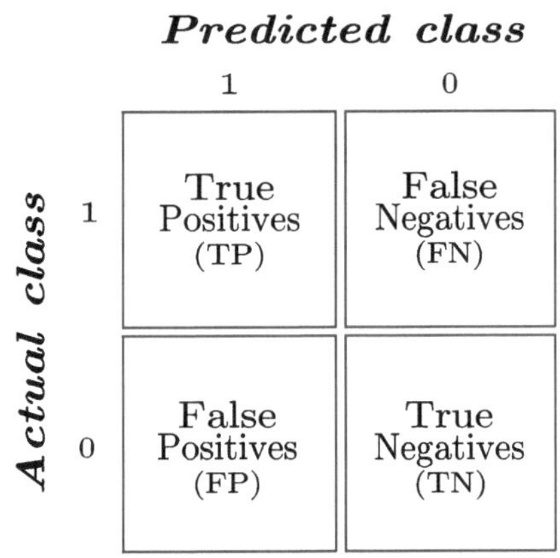

Fig. 3.2 A standard confusion matrix for a binary classification problem. The cells show the number (or the percentage) of samples that meet the characteristics indicated by the rows and columns

- True Negatives (TN): the number of times the model does not predict the condition when the condition is absent;
- False Positives (FP): also known as "Type I error", the number of times the model predicts the condition when the condition is absent;
- False Negatives (FN): also known as "Type II error", the number of times the model does not predict the condition when the condition is present.

The following features can be defined as well: (i) the total number of actual negatives (TN + FP); (ii) the total number of actual positives (TP + FN); (iii) the total number of predicted negatives (TN + FN); (iv) the total number of predicted positives (TP + FP); (v) some reference metrics.[1]

Accuracy—One of the most used metrics, it ranges from 0 to 1 and represents the ratio of correctly predicted observations to the total ones.

$$A = \frac{TP + TN}{TP + FP + FN + TN} \tag{3.3}$$

Instinctively assumed as a great measure of the model performance, it actually provides very little information about false positives and false negatives. In general, it is not recommended to base the assessment of the goodness of a classifier model solely on Accuracy since it is a metric that can easily mislead. Indeed, other metrics, such as Precision and Recall, can assess the impact of those wrong predictions on the model performance.

Precision—Also known as Positive Predictive Value (PPV), it ranges from 0 to 1 and is the ratio of true positives over all the true and false positives:

[1] Many metrics can be used to measure the performance of a classifier: Accuracy, Precision, Recall, Specificity, F Measure (F1,F0.5,F2), Matthew's Correlation Coefficient, ROC Area, Fallout. Only Accuracy, Precision, and Recall will be analyzed in detail as those mainly used in the continuation of this thesis.

$$P = \frac{TP}{TP + FP} \tag{3.4}$$

High precision means that the model's total positive predictions are mainly composed of cases where the condition is present (correct guesses); it can be seen as a percentage error on the total number of predicted positives.

Recall—Also known as True Positive Rate (TPR), it ranges from 0 to 1 and represents the ratio of true positives over the sum of true positives and false negatives:

$$R = \frac{TP}{TP + FN} \tag{3.5}$$

High recall means that the model correctly predicts most of the cases where the condition is present; it can be seen as a percentage error on the total number of actual positives.

The classification theory has developed over the years different classifier models, from the most straightforward and lightest—computationally speaking—to the most complex and most refined. Whether and when to apply one or the other is strictly dependent on the type of application. Some of the most common classification algorithms used in ML problems are reported below [13].

Dummy Classifier—Also known as the "Baseline classifier", it gives a baseline accuracy (also called "null rate") that shall always be checked before selecting a more sophisticated classifier. A dummy one is a classifier whose behavior, rather than fitting a model by generating insight into the training data, relies on simple rules to perform the classification task. It is also used as a comparison tool to verify that there are no errors in implementing other types of classifiers since any different classifier is expected to perform better on the given dataset since any analytic approach for a classification problem should be better than a random guessing approach. Among the possible classification criteria, the most used are: (i) "Most Frequent", which makes the dummy classifier always predicts the most frequent class label in the training data; (ii) "Stratified", which generates probabilistic predictions by respecting the class distribution of the training data; (iii) "Uniform", which makes predictions uniformly at random; (iv) "Constant", which always predicts a constant class label.

Logistic Regression (LR)—Logistic Regression is a simple but powerful classification algorithm and, for these reasons, one of the most exploited to tackle binary classification problems. Nonetheless, there are also extensions to deal with multi-class scenarios. In the One-vs-Rest (OvR) logistic regression, a separate model is trained for each class, predicting whether an observation corresponds to that class or not, and assuming that each classification problem is independent. Logistic regression is named after the function used at the core of the method: the logistic function, also called the Sigmoid function. The output of the Sigmoid function is interpreted as the probability of a specific sample to belong to one of the two classes given its input features x_i linearly combined by the weights w_i. This probability is then processed by a quantizer that sets a minimum probability threshold of belonging to the given class, returns a zero (i.e., the sample does not belong to the class) or a one (i.e., the sample belongs to the class). The training phase of this classifier is achieved by modifying the weights w_i through a process that optimizes the likelihood function [14].

Support Vector Machine (SVM)—Support Vector Machines are classifiers primarily used to address supervised learning problems. Their popularity is due to their ease of being adapted to both linear and non-linear cases. In linear problems, they aim to define the best "decision hyperplane" that separates the involved classes by maximizing the distance between this hyperplane and the closest sample points, called support vectors. Two are the ways used to handle non-linearly separable datasets: (i) a slack variable is introduced to relax the linear constraints and to allow for the optimization process convergence; (ii) in the kernel-based methods, samples are mapped onto a higher dimensional space to make them linearly separable; the decision hyperplane can thus be detected and then the inverse transformation applied to project it back to the original space [15, 16].

Decision Trees (DT)—Decision Trees classifiers rely on dividing the dataset by asking if-questions until it is narrowed enough to make a prediction; consequently, their graphic representation looks like a tree ramification. This kind of classifier is still used in supervised During the training and validation phase, these algorithms learn the most suitable series of questions to correctly predict the input data class. The maximum depth (i.e., the maximum number of branches from the roots to the leaves), the maximum and the minimum sample number for each branch, and the maximum feature number to check at every split represent the most relevant hyperparameters to be determined. When many features characterize the dataset, they usually tend to overfit, remarkably deteriorating their generalization performance. In such cases, the results of multiple decision trees, developed around the same dataset but considering less randomly chosen features, are combined: they are called Random Forest (RF) algorithms [17, 18].

K-Nearest Neighbors (KNN)—In supervised learning, K-Nearest Neighbors (KNN) classifiers are part of the so-called lazy algorithms since the dataset is memorized rather than learned to be distinguished. Once defined an integer K and a distance metric (e.g., Euclidean distance, Manhattan distance), they classify each new sample by finding the other K closest samples and assigning to the new one the more numerous label in the neighborhood. The main strengths of the lazy algorithms, as KNN, are that they do not require a training phase and that they adapt significantly well to new data. Conversely, the major drawback is that the computational cost of the classification task increases every time a new data is labeled since more distances have to be calculated [19, 20].

3.2 Embedded Hardware

At the beginning of this thesis work, the exoskeleton prototype required substantial modifications to its electronic apparatus. The button control system and the absence of sensors for finger position feedback imposed strict limits on the device's usability.

Moreover, the simultaneous drive of four motors for the reproduction of the same movement on all the four long fingers was challenging to manage and particularly subject to synchronization issues. The main components that replaced the previous electronics are reported here below in Fig. 3.3.

The choice of these components has been crucial for setting up the new embedded hardware. Having adequate performance was not the only constraint that guided the pick. Each part has been selected in agreement with the guidelines of a research activity aimed at developing a control strategy for a low-cost wearable device for clinical use: it should have been small, lightweight, low-powered, easy to set up, and as cheap as possible.

The design of the new hardware started with the choice of the microcontroller. It is the component that has both computational power and storage capacity thanks to the integrated microprocessor and memory; besides, it is capable of interacting with different input/output peripherals.

In light of these reasons, it is good practice to choose the microcontroller first because it is the nerve center of the control system: not only all the other components will have to interface with it, but also its performances will determine the general complexity of the manageable hardware architecture and the computational weight of the executable control logic.

Distinctive features of a microprocessor are usually its architecture (e.g., AVR, ARM), the clock frequency, and the bus width—which determines the number of bits that can be processed per computational cycle. The integrated memory is instead

Fig. 3.3 The main electronic components used during the device revamping process: from the top left corner, a servomotor, a microcontroller board, a Bluetooth module, two sEMG sensors, a motor driver and a custom PCB

Fig. 3.4 The Arduino Nano board and the HC-05 Bluetooth module

evaluated by the number of bytes that can be stored inside the Flash memory—used to store the code—then, the SRAM memory—used for managing dynamic variables—and, finally, the EEPROM memory—used to store long-term information. Then there is additional information that should be considered for the appropriate choice of the microcontroller: power supply voltage, energy consumption, logic level, number of input and output ports, embedded ADCs, communication protocols with other devices, and support for wireless connection.

The first-stage exoskeleton exploited an Arduino Nano controller board from the Arduino company (Fig. 3.4). The characteristics of such a microcontroller board—the main ones will be listed right below—have been considered sufficient for implementing the new control strategy and, therefore. Therefore, it has been confirmed as part of the project. It is based on the microcontroller ATmega328p from Atmel (part of Microchip Technology Inc. since 2016), which is its computational and storage core: 8-bit AVR architecture, 16 MHz clock frequency, 32kB, 2kB and 1kB of, respectively, Flash, SRAM, and EEPROM memory. The tiny Printed Circuit Board (PCB), in just 18×45 mm and 7 g, offers a Universal Serial Bus (USB) programming interface and ready-to-solder pins for connecting to the 32 ports of the ATmega328p. Of these 32 lines, 23 are general purpose input/output, while the others are reserved for the serial, SPI, and I2C communications, and the power supply. Even if the Atmega328p has an operating voltage of 5 V, the board accepts from 7 to 12 V as input supply, thanks to the embedded voltage regulator, and has a power consumption of about 20 mA. The presence of a 10-bit 6-channel ADC and the possibility of generating up to six PWM signals make this board perfect for quick interfacing with different types of sensors and actuators. The cost of the Arduino Nano is about 20 €.

Figure 3.4 shows, alongside the microcontroller board, the Bluetooth module HC-05 from Guangzhou HC Information Technology Company. It is one of the most popular and inexpensive class-2 Bluetooth modules used for wireless communications. Its cost is less than 10 Since it can be found already mounted on a breakout board, it is easy to implement in projects that involve microcontroller boards. This module has a range of 10 m, its communication parameters can be changed through AT commands, and it is programmable both as a master and a slave device—the

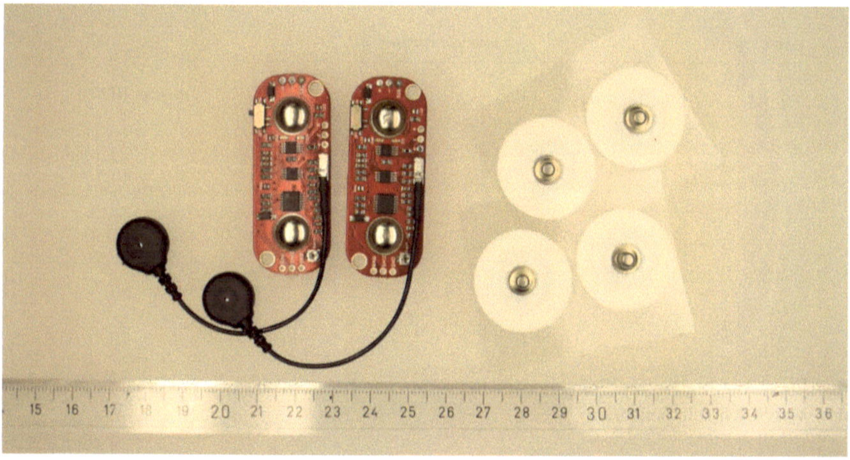

Fig. 3.5 Two MyoWare™ Muscle Sensors and the Ag/AgCl snap electrodes used to capture the sEMG signal

former is the one that looks for connections, the latter instead waits for an external connection to be received. The supply voltage can be between 3.6 and 6 V, the logic level is at 3.3 V, and the size and the weight of the component are, respectively, 28 × 15 mm and 5 g. The current required in the coupling phase is about 35 mA, while during the communication, it drops to about 8 mA. The HC-05 turn a serial port into a Bluetooth 2.0 one, generally with a Serial Port Profile (SPP), thus becoming a serial-over-Bluetooth communication port with a configurable Baud rate up to 1382400 bps. This component has been added to the HES to provide it with a wireless connection that could be used to manage the sEMG collection from a computer and remotely monitor the exoskeleton's functioning.

Confirmed the microcontroller, the choice of the other components has focused on those with tested and reliable compatibility with it. This research becomes problematic in the field of surface electromyographic sensors, as many types are intended for exclusive use in clinical or laboratory environments rather than on embedded platforms. Several other factors are also not negligible. They are usually sold accompanied by dedicated software that allows the management of the collection characteristics. Such software generally runs on a desktop operating system, Microsoft Windows in most cases, and is distributed under license. Besides, the source code is often not released and is the only way to intervene in data sampling. Finally, it is worth noting that the cost of these systems is around several thousand euros, which is not compatible with the low-cost goal of this research activity. All of these reasons drove the research elsewhere.

A suitable solution for the sEMG signals collection has been found in the MyoWare™ Muscle Sensors from Advancer Technologies (Fig. 3.5). These sensors incorporate the housing for two monopolar snap electrodes into a small breakout board (20.8 × 52.3 mm) specifically designed to work with low-powered develop-

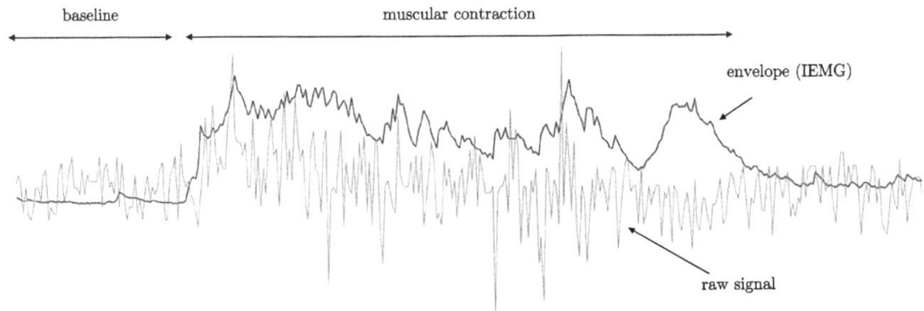

Fig. 3.6 A qualitative representation of the difference between the two possible outputs of the MyoWare™ Muscle Sensors

ment boards like the Arduino and the Raspberry Pi. Even if there are two electrodes, the MyoWare™ behaves as a single bipolar sensor capable of generating two different types of differential output. The third electrode housing is thought to host the reference electrode. Since the computational power available on microcontroller boards is notoriously low, these sensors have been equipped with an internal control logic that partly preprocesses the signal before outputting it. In such a manner, it is possible to have as output the raw sEMG signal and the envelope of the rectified signal (see Fig. 3.6). This feature is the equivalent to the IEMG—introduced in Table 3.1—and already gives an idea of the force associated with the muscle contraction. The formula for the IEMG calculation is given below:

$$IEMG = \sum_{k=1}^{N_e} |x_k| \tag{3.6}$$

where x_k is the k-th sample of the time window on which this parameter is calculated, and N_e is the total number of samples of the same time window.

These measures are carried by a single wire and can be read by an analog input port. They can be powered with a voltage from 2.9 up to 5.7 V and have a maximum power consumption of 15 mA The necessary power can be supplied directly from the output port of any microcontroller board. With a straightforward design and a low price (~35 € each), they are an excellent option to go for low-cost wearable applications. The choice of using only two of these sensors will be justified in the next section.

Likewise, the adoption of an actuation system equipped with a single servomotor will be discussed in Sect. 3.3. The servomotor in question is the HS-5495BH High-Torque Servo from Hitec, visible on the left in Fig. 3.7, next to the SuperModified V3.0 driver from 01™ Mechatronics. The possibility of integrating such a motor driver was discriminating in the actuator type and model choice.

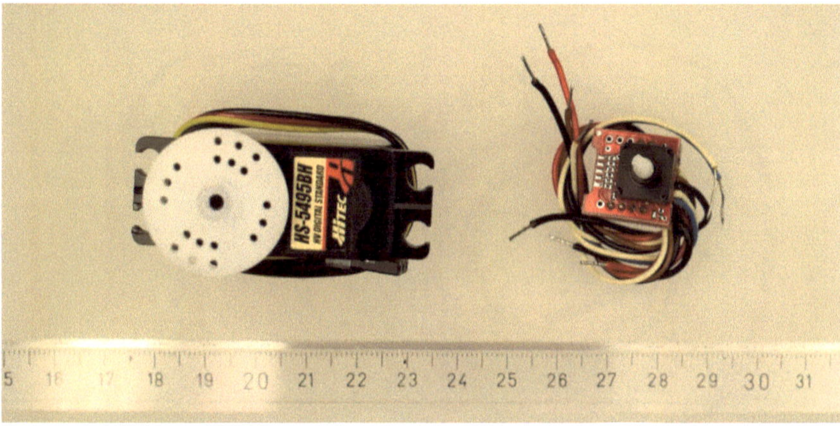

Fig. 3.7 The HS-5495BH and the 01™ Supermodified V3.0 driver

The HS-5495BH can be powered with a 7.4 V power supply (the capability of a standard 2-cell LiPo battery) with a power consumption of 190 mA in non-load operations and a maximum stall current of 1400 mA. The maximum output torque is about 0.7 Nm, while the maximum achievable angular speed is 6.7 rad/s. These features have been heuristically and empirically considered sufficient for the movement of all four long fingers. In fact, this motor had to replace the four old actuators in the new electronic setup. Its dimensions, weight, and cost are $39.8 \times 19.8 \times 38$ mm, 45 g, and 25 €.

The 01™ Supermodified V3.0 is a fully enclosed motion control module composed of three stand-alone electronics layers: (i) a stage that manages the correct power supply (from 5 up to 24 V and up to 5 A) of the other two stages and to the motor; (ii) a controller stage, also equipped with the ATmega328p microcontroller, wholly dedicated to delivering a closed-loop PID control at 9.765 kHz on the angular position and speed of the crankshaft; (iii) a sensitized stage provided with a 15-bit absolute magnetic encoder that gives angular feedback with 32768 discrete positions per revolution. The communication with the 01™ Supermodified V3.0 is straightforwardly achieved through standard protocols as the RS-485, UART, SPI, or I2C. Besides, it is possible to fit this motor driver inside the HS-5495BH replacing its original electronics. In just $16 \times 16 \times 27.5$ mm and for a price of 70 €, it turns a servomotor into a fully controlled—position and speed—continuous rotating motor directly interfaceable with the Arduino Nano board.

From the tests conducted on the previous version of the exoskeleton arose, as reported at the end of Section 2.3, that a good part of the usage safety deficiencies was due to the total absence of automatic control on the fingers' position. The 01™ MagEnc V3.0 Low Rev magnetic encoder—the same encoder contained in the 01™ Supermodified V3.0—has been added to the index finger mechanism to compensate for this lack. In particular, concerning Fig. 2.1, the encoder was chosen to be mounted on the axis of Joint ① and integral with the component C to measure the

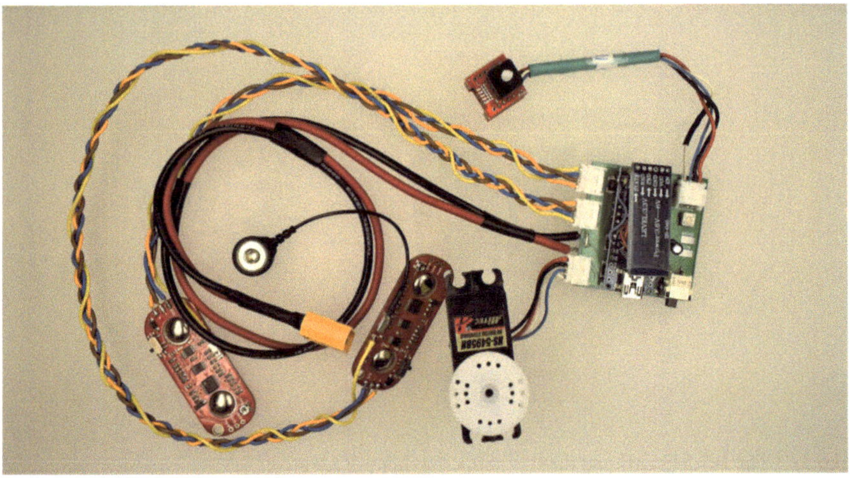

Fig. 3.8 The new HES embedded electronics

control variable α_2 and, consequently, to provide an indirect measurement of the MCP rotation. This sensor can also measure the angular speed directly. The index finger mechanism has been chosen as it offers more lateral space available for the 01™ MagEnc V3.0 Low Rev installation Since all four long fingers are meant to be moved simultaneously, the index measurements are assumed representative for the status of the other fingers.

The final step of this process was designing a custom PCB to house all the components exploiting the Electronic Design Automation (EDA) software Autodesk EAGLE. The result is visible in Fig. 3.8, which shows how the whole electronics has been hooked up together to the PCB. As already specified, the new electronic hardware still has at its core the microcontroller board Arduino Nano. Since it is already tiny and dimensionally optimized, it has been opted to house the whole board on top of the PCB. In addition to the one reserved for the Arduino Nano, the space for four JST connectors (4 pins, 2 mm pitch), a bistable switch, an RGB LED, and a circuit both dedicated to monitoring the status of the battery have been set up on the new custom board. The use of JST connectors—reserved for the motor, the two sEMG sensors, and the encoder—has been preferred to the direct welding of the components on the board to facilitate their eventual replacement. It is powered using a standard 2-cell LiPo battery (7.4 V), and the routing has been made using two different track measures. The thinner tracks connect the components with low current consumption—such as the encoder and the sEMG sensors—and have been oversized for safety with a thickness of 0.4 mm for current transport up to 1 A. The thicker traces have instead been destined to the motor power supply and dimensioned for a current flow up to 2 A, 40% greater than the current absorbed by the motor in stall conditions. The final dimensions of the board are $45 \times 45 \times 2$ mm.

3.2.1 sEMG Bracelet

Some issues regarding the wiring management and the dressing of the electromyographic sensors have emerged during the preliminary assessment of the new integrated electronics. Such tests have been performed in the laboratory and involved only healthy subjects. The experiments have been carried out by asking the subjects involved to wear two MyoWare™ sensors on the forearm and trigger the motion of an HS-5495BH servomotor with the muscle contraction. The sEMG sensors were worn on the *extensor digitorum* and the *flexor digitorum*. The control logic was based on the signal amplitude thresholds presented by Ochoa et al., reported in Sect. 1.3. The subjects were sitting in front of a table on which the servomotor was fixed.

The motor motion was regularly activated according to the muscular contractions of the subjects. However, the wrist movements could sometimes establish excessive tensions in the cables that connected the sEMG sensors to the microcontroller. These stresses quickly caused the detachment of the electrodes from the skin, compromising the signal acquisition. In an attempt to solve this problem, a first-tentative forearm bracelet for the sEMG collection has been designed (see Fig. 3.9). The bracelet integrates some of the components already identified, such as the microcontroller, the Myoware™ sensors, and the Bluetooth module, with a 500 mAh Li-ion battery, a custom made 3D-printed case, and a Velcro band. This solution has proved to be valid for decoupling the acquisition system from the movements of the wrist (thanks to the Bluetooth bridge for data transmission) and lightening the workload on the original processor (taking care of the sampling and eventual preprocessing of the sEMG signals).

Fig. 3.9 The sEMG bracelet. The central unit houses the microcontroller, the Bluetooth module and the battery. The two lateral units contain the two Myoware™ sensors. The total cost and weight are about 150 € and 80 g

3.3 The Novel Control Strategy

The lack of sufficient safety for the user and the limited intuitiveness of the original prototype's use have strongly highlighted the necessity to develop an alternative control strategy. Both issues have arisen in the light of the first clinical trials performed on a single patient and were mainly due to the absence of automatic control over the position of the fingers and the uncomfortable use of a two-button actuation strategy.

The scarce safety of the fingers' range of motion control has been solved by inserting the feedback of an angular sensor in the new control strategy. The magnetic encoder presented in the previous section, has been integrated into the index finger mechanism to monitor the position and angular speed of the controlled joint. The details of the control strategy implemented thanks to the angular feedback from this sensor will be discussed in Subsection 3.3.1, where the overall architecture of the new controller algorithm will be examined and analyzed. The second issue concerned instead the difficulty in actuating the HES. The test user—a man born with SMA—pointed out that, once the exoskeleton was worn on one hand, having to use the other to operate the device by pressing the buttons was very uncomfortable and more cumbersome than not using the exoskeleton. As a matter of fact, the user had developed over time alternative strategies of object grasping and handling for which the reduced residual mobility of his hands was sufficient. From this point of view, it was necessary to provide him with a device that would at least allow him to independently control the movement of the hands so that its effectiveness could then be compared to that of the replacement techniques he developed.

Finally, it is essential to remember the choice to focus on developing a real-time strategy for the classification of the user's motor intentions rather than of a motion control technique. The mechanical structure of the device, already optimized to accurately track the natural trajectory of the fingers, did not require special attention indeed in controlling position and speed, except those just mentioned. Once the user's intention has been recognized, the device would be triggered and actuated, relying on the PID control already integrated into the motor.

3.3.1 Control Algorithm

The control algorithm proposed in this thesis is born for managing the motion of an assistive hand exoskeleton meant to help people with clenched fist deformity open the hands [21–23]. This device consists of a motor block positioned on the back of the hand—containing the actuation systems and the control electronics—and four planar 1-DOF mechanisms placed above the long fingers. The four finger mechanisms are optimized at the mechanical design stage to optimally track the finger trajectory of the specific patient for whom the device is designed. Their movement derives from the action of cables that, when wound or unwound by servomotors, pull the mechanisms in the extension or accompany their flexion—no forces are transferred to the fingers during flexion. In the original version of the control system, the actuation command was triggered using two buttons pressed by the user. The same buttons also managed the ROM of the fingers: the user himself released the actuation button once he reached limits he deemed right. However, such a preliminary technique resulted in being remarkably uncomfortable and unsafe.

The basic idea for the new control strategy implementation relies on introducing new sensors and using a different control algorithm to innovatively manage the information provided by the new sensor technology. Such information will be needed to manage the fingers' ROM predict the motor intentions of the user without having him perform a physical activity that involves the use of the hands. In addition to the above-reported features, the resulting overall system still requires meeting the requirements introduced in Chapter 1 (i.e., to be fully wearable, lightweight, low-powered, stand-alone, low-cost). In light of these considerations, microcontroller constituted the most suitable hardware technology for developing the new control strategy by representing a feasible trade-off for the long list of design patterns and requirements. Hence the choice of all the devices detailed in Sect. 3.2. However, this kind of hardware offers limited resources and computational power compared to desktop PCs, workstations, or industrial controllers. Therefore, the design of the new control system has begun with the awareness that a compromise would have to be found between the performance of the refined techniques tested on benchtop computers and the computational lightness required by the used hardware.

As already pointed out in Sect. 1.3 and reiterated at the beginning of this chapter, current literature reports more and more convincing studies on the application of control techniques for wearable robotic devices based on surface electromyographic signals. These methods propose to classify the person's motor intention within a set of possible classes and, consequently, control the robotic system by activating the state corresponding to the recognized class. Since sEMG signals are not straightforwardly decipherable, most of these strategies use the high computational power distinctive of those bench equipment that is not suitable for wearable applications. There are also cases in the literature where naive variants of these methodologies have been implemented on wearable devices. The trade-off made in all these control strategies concerns the maximum number of classifiable movements. This design pattern is explained by taking into account that increasing the number of classes to be identified unavoidably increases, in turn, the complexity as well as the computational cost of the algorithm in charge of fulfilling the classification task. Algorithms with a high computational cost require a longer time to be executed—time which extends as the performances offered by the processors decrease—and, therefore, they risk resulting unresponsive and not functional. The literature affirms that a control latency higher than 300 ms is perceived as a considerable delay and produces an intense frustration in the user [24].

When it comes to exoskeletons for the hand, this aspect becomes even more complicated. First of all, the limited space available around the upper limb and the low bearable weight means that the exoskeletal system shall be considerably small. Consequently, by minimizing the size, the volume of the electronics integrated into such a device is reduced; generally speaking, its computational power is also diminished. Finally, this limited technological scenario has to deal with the very high natural dexterity that a healthy hand would possess. The human hand can indeed perform lots of different movements, which are products of the contraction of several muscle bands located very close to each other inside the forearm. In light of what is reported in Sect. 3.1.1 about the nature and the noise level of the sEMG signals, being able to accurately classify all the possible movements of the upper limb from these signals necessarily requires the use of high computational power machines running heavy ML algorithms—and yet, nowadays, it is still not possible to distinguish them robustly.

In this complex scenario, a suggestion for an alternative approach comes from a recent study, which demonstrates that the independent use of the fingers is not crucial in carrying out most of the common Activities of Daily Living (ADLs) [25]. From this point of view, it is possible to think of an assistive HES that could be of practical use even without allowing the movement of individual fingers. This assumption, which in itself dramatically reduces the number of hand gestures to be distinguished, also has a positive impact on the architecture of the mechanical and electronic hardware. The choice on the number of electronic components reported in Sect. 3.2 has been made considering classifiable just the three main motions of the hand, i.e., opening, closing, and resting. The use of only two MyoWare™ Muscle Sensors is the result of the combination of two facts: firstly, two is the minimum number of muscle bands whose myoelectric activity has to be recorded to classify these three movements; secondly, a low number of sensors is preferable in a low-cost perspective and aiming at high wearability. A single motor is required for the actuation system as it is possible to move all the fingers together. Besides, remembering that the actuation is cable-driven, a pulley with four different diameters applied to the crankshaft shall allow the fingers' synchronous motion despite their different size. Finally, a single encoder is sufficient to characterize the motion of all fingers once ascertained that the correct sizing of the multi-diameter pulley of the motor guarantees the same angular velocity for them all.

Figure 3.10 and Algorithm 3.1 show the general scheme of the new control strategy, which can be split into two sequential subroutines: the "classification loop" and the "actuation loop". The first one is performed on the microcontroller integrated into the data collection bracelet (Sect. 3.2.1). In contrast, the second one runs on the microcontroller placed inside the motor block and is responsible for managing the device actuation. Everything is designed to be run at a frequency 50 Hz, considered a suitable control frequency for a system, like the muscular one, interested by slow dynamics—and even slower when it comes to patients suffering from motor disabilities. The control strategy will now be described by starting from the quick analysis of the subroutine that, downstream of the classification loop, handles the actuation of the exoskeleton, as it is conceptually easier and composed of only standard procedures. More space will be left to the other loop description since it is the real innovative point compared to the other control techniques for HES based on the classification of electromyographic patterns.

Algorithm 3.1 Pseudo code of the proposed control strategy

```
Input:
F_p = current finger position
F_v = current finger velocity
M_a = current muscular activity

Classification loop:
if M_a compatible with opening gesture then
 | return open
end
else if M_a compatible with closing gesture then
 | return close
end
else
 | return rest
end
Actuation loop:
if open then
    if F_p compatible with range of motion then
     | run motor to pull the cables to assist the hand opening
    end
    else
     | max opening reached: stop motor and wait for a new close command
    end
end
if close then
    if F_p compatible with range of motion then
        run motor to release the cables to follow the hand closing
        if F_v lower than velocity threshold and motor is running then
         | object grasped: stop motor and wait for a new open command
        end
    end
    else
     | max closure reached: stop motor and wait for a new open command
    end
end
if rest then
 | do nothing: wait for a new different command
end
```

All that is required of the actuation loop can be summarized in the translation of the classification loop output (namely an opening, a closing or a resting intention) into a set of valid low-level commands to be sent to the actuator, while two inner loops take care of controlling the ROM of the fingers and the eventual grasping of an object.

Every time an opening intention is detected, the motor will be requested to spin in order to pull the cables connected to every finger mechanism. Therefore, the result will be an opening movement of the fingers, which shall take place as close in time as possible with the user's muscle contraction. Similarly, the motor shall spin in the opposite direction as soon as a closing intention is classified. Besides, in this case,

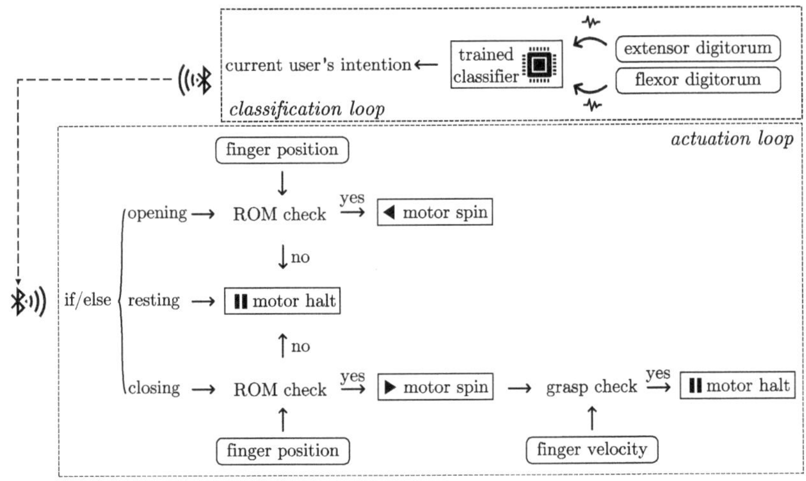

Fig. 3.10 The block diagram of the overall control strategy. In the top box, the classification loop that, given the two signals sEMG, returns the user's intention. In the bottom box, the actuation loop that, given the user's intention, handles the actuation system of the exoskeleton. The motor commands (identified by ▮▮, ◀ and ▶) are executed until the next one is received

the motion of the exoskeleton should take place without excessive delay from the muscle contraction. When the classification loop recognizes neither of the previous two intentions, the actuator will be required to hold its position. Commands are given through the 01™ Mechatronics proprietary libraries provided with the motor driver.

The actuation command is sent to the motor at a frequency 50 Hz to ensure that the exoskeleton's motion is perceived as sufficiently reactive by the user. The exoskeleton then reproduces a different movement up to 50 times per second; subsequent intentions belonging to the same class are translated into a single prolonged actuation. This design choice presents a further benefit: complete control over all the intermediate positions of the ROM. In fact, unlike the strategies based on motion triggers, the user can interrupt the motion of the exoskeleton, at every time once reached the desired degree of opening/closing, by simply relaxing the corresponding muscle contraction. However, the need to maintain a stable level of muscle contraction to achieve a smooth HES motion emerges as an inevitable price to pay; the device control will then be more fatiguing in the long run.

As previously mentioned, the translation of the user's motor intention into actual exoskeleton movement is not the only thing handled by this control loop. At each control cycle, the angular position of the fingers is measured before the actual actuation takes place. The MCP joint position feedback is real-time measured using the magnetic encoder mounted on the index finger mechanism. This measure is checked before issuing the command to the motor to verify that the system lies within the permissible ROM identified during a preliminary phase. A physiotherapist, firstly, makes the user wear the device and, secondly, manually moves the patient's hand, making complete opening and closing gestures several times. During this process, the exoskeleton control system records the maximum and the minimum value measured by the encoder and set them as the admissible ROM upper and lower boundaries. The ROM upper and lower bounds are then extracted from the average of the maximum and minimum values recorded by the encoder. Subsequently, if the encoder records that one of these limits has been reached during the regular functioning, the motion is immediately stopped. The entire system waits then to classify the correct command to restart moving in the opposite direction. E.g., once the maximum closing limit has been reached, a halt is sent to the motor, and any other closing command is rejected; the motor will move again only after an opening command. Therefore, the device's poor safety problem resulting from the absence of automatic control over the admissible ROM has been therefore solved.

While the outermost control loop supervises the ROM, a nested loop is in charge of detecting the grasping of objects while closing the hand. Since there are no force sensors on the HES, this detection is performed comparing the angular velocity measurements of the MCP joint and the motor crankshaft. When the first one drops over a certain threshold while the motor still runs, continuing to unroll the cable, it is reasonable to think that the fingers have encountered an object or an obstacle. In these cases, the control system commands the actuator to stop before an excessive cable length is released. Preventing excessive cable unwinding by actively controlling the grip of objects has proved necessary since the first laboratory tests. Before this additional control level was introduced, the user had to relax the flexor muscles to stop the motor motion while grasping an object.[2] Nevertheless, the transition from closing to resting state has proved to be challenging when interfacing with an object, often making the motor run after the actual grasping. As a result, the cable kept on being released even when not requested: this risked giving rise to cable twisting and introduced a deleterious delay in control actions after the grasping.

The subroutine in charge of taking care of the sEMG signal classification is the first to be executed. The two Myoware[MT] collect every 20 ms the muscular activity of the *extensor digitorum* and the *flexor digitorum*. The sEMG signals are then processed by the Arduino Board on the bracelet and classified into one of the three reproducible user intentions: hand opening, closing, or resting. The EMG envelope has been preferred over the raw output for two different reasons: (i) to make most of the electronics integrated on the sensors and not to burden the microcontroller ADC by sampling the raw signals; (ii) because this feature provides a qualitative idea of the state of the muscle activation, and it has proved to be convenient for the discrimination of the three gestures mentioned above. Once the current user's intention has been classified, the corresponding signal is passed to the actuation loop that translates it into the appropriate control commands for the actuation system.

[2] As much as having to relax the muscles to maintain the grip on an object may seem counter-intuitive; it is crucial to remember that the exoskeleton has been designed for people whose hands are forced to fist by a considerable tendon tension and, therefore, can exert passive closing forces even when their muscles are relaxed.

The related works reported in Sect. 1.3 present similar procedures: first, they independently monitor the myoelectric state of the same muscle bands and, second, they recognize the user's intention by comparing the signal amplitude with respective lower thresholds corresponding to the minimum levels of muscle activation for some given movements. As a consequence, these strategies make the classification process one-dimensional. However, even though this method is computationally very light and produces a decent control experience, it carries some inherent problems. Unwanted movements, minor impacts with objects in proximity of the sensors, and skin-electrode sliding might indeed produce motion artifacts. Although not corresponding to actual muscle activation, these artifacts can easily be misinterpreted by the sensors to which they appear as muscular activity, generally of high intensity. Besides, different movements might result in the same muscular activation because of the cross-talk effect. Wrist motion when handling objects is a typical example of how this problem can be harmful for such an application since the wrist's extensor muscles are located under the fingers' ones. An extension movement of the wrist would indeed most likely activate the opening motion of the exoskeleton, producing the hand opening and the consequent fall of the object. Being subject to these types of disturbance makes these methods considerably vulnerable and prone to misclassifications.

These unforeseen events may cause, in fact, damage to the system or even to the user. Aiming at reducing such a risk, the new control strategy adopted a different classification criterion, which no longer relies on the value of the individual sensors but on the combination of the measures provided by both. Rather than independently comparing the two signals to a lower threshold, every sample interval, a point—whose coordinates are the envelopes of both the sEMG signals—is plotted on a 2D Cartesian plane (as visible in Fig. 3.11) and is tested against the belonging to a specific area. Thus, the classification process becomes two-dimensional, and a disturbance acting on a single muscle has fewer chances to lead to misclassification.

The choice of the classifier required more than one attempt to find an optimal solution for the under-consideration case and resulted in an iterative procedure rather than a deterministic process. However, the procedure followed for each model remained unchanged and was characterized by the steps below: (i) a sufficient dataset has been collected; (ii) the original dataset has been split into the training and validation set, equivalent to the 80% of the original one, and the testing set, corresponding to the remaining 20%; (iii) accuracy, precision, and recall have been selected as the evaluation metrics; (iv) during the training and validation phase, the cross-validation and the grid search techniques have been exploited to tune the classifier parameters and hyperparameters; (v) the test dataset has been used to extract the real metrics of the trained model. The whole process has been carried out using the open-source library Scikit-learn: the gold standard, in terms of ease to use, flexibility, and completeness for Machine Learning in Python [26]. All the different algorithms introduced in Sect. 3.1.2 have been trained on the randomly chosen 80% of the dataset shown in Fig. 3.11 to distinguish the user's intentions on a set consisting of only two classes: hand opening and closing. The resulting metrics for the different classifiers are visible

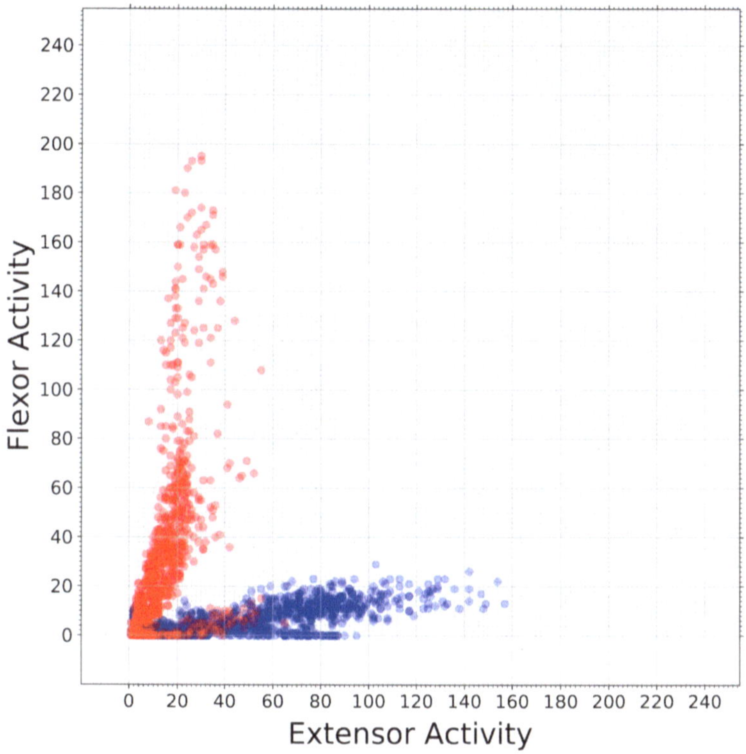

Fig. 3.11 The graph shows a two-dimensional representation of the forearm muscular activity during hand opening (blue) and closing (red). On the abscissa axis the sEMG signal measured by the sensor placed on the *extensor digitorum* is shown, whereas on the ordinate axis the signal from the *flexor digitorum* is reported. The scale of the axes will be discussed in Sect. 3.3.2

in Figs. 3.12 and 3.13, where the confusion matrices and the corresponding decision boundaries are reported.

Three are the principal considerations that can be done by analyzing the results reported above. Firstly, by looking at the different accuracy, precision, and recall values, it can be seen that the Dummy classifier arises by far as the worst-performing algorithm among the tested ones. It is relevant to notice that the verification of such a condition provides a preliminary (but essential) confirmation of the classification process's goodness. Secondly, continuing to focus on the perspective of quantitative analysis in terms of the judgment metrics, SVM—in the radial-basis-function-kernel-based version—and KNN highlighted outperforming results by best fitting this non-linearly separable dataset. Thirdly, none of the algorithms achieved an acceptable level of accuracy, precision, or recall.

This last datum can be interpreted by concentrating the attention on the decision boundaries shown in Fig. 3.13. In qualitative terms, by merely looking at how the two classes are divided in the Cartesian plane, it can be noted that the classification

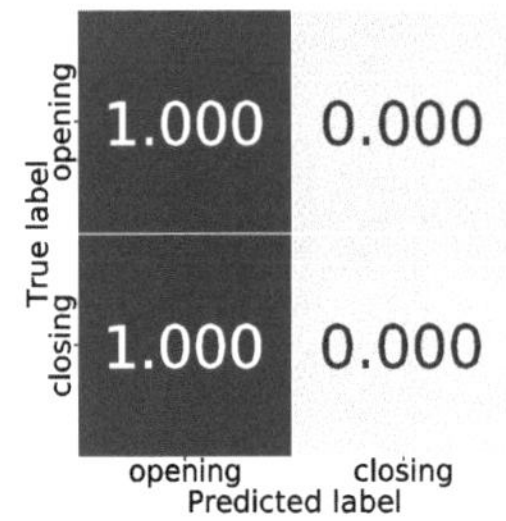

(a) Dummy classifier
A: 0.557, P: 0.278, R: 0.500

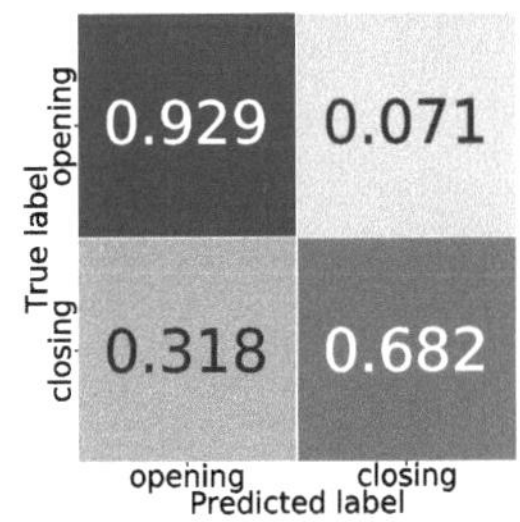

(b) LR classifier
A: 0.820, P: 0.835, R: 0.805

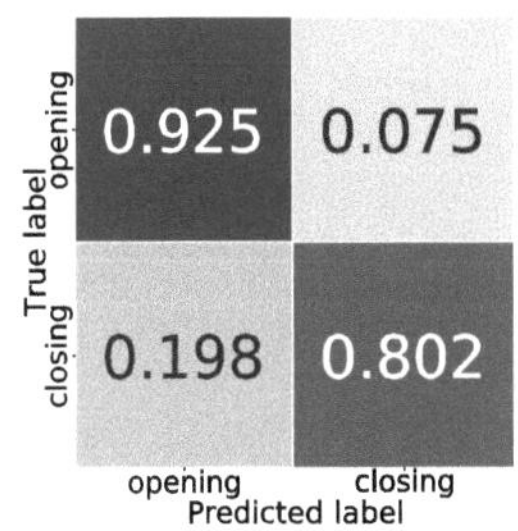

(c) SVM classifier
A: 0.870, P: 0.874, R: 0.863

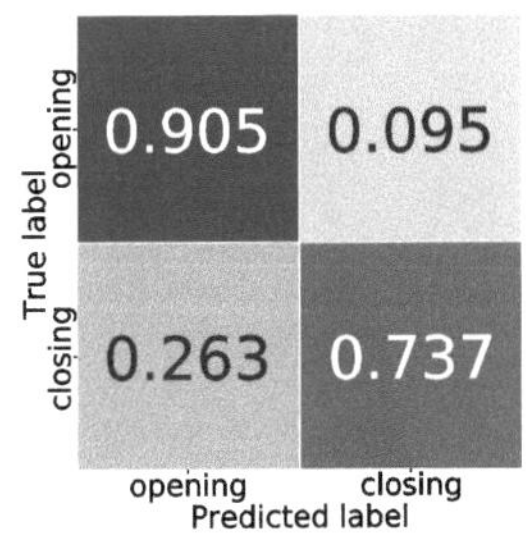

(d) DT classifier
A: 0.828, P: 0.835, R: 0.821

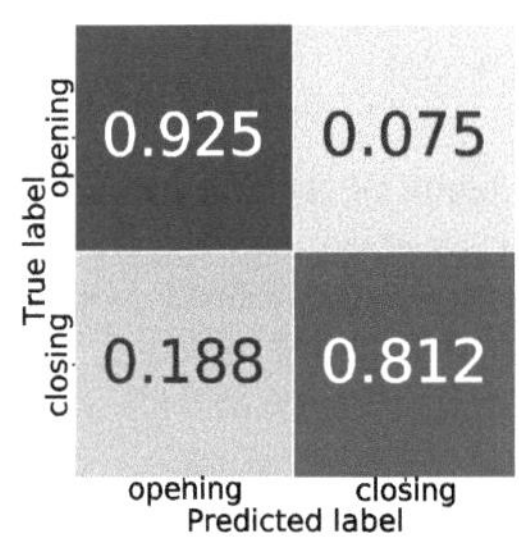

(e) KNN classifier
A: 0.878, P: 0.880, R: 0.869

Fig. 3.12 The confusion matrices from the testing of the five 2-class classifiers. A stands for Accuracy, P for Precision, R for Recall

was very accurate in those areas which were representative of a substantial muscle contraction—i.e., where the flexor or the extensor activity signal assumed large values. However, the same classification accuracy could not be achieved where the muscle contraction was weak, and the corresponding sample fell around the origin. As the graphs show, the sEMG signals collected during the opening and closing of the fingers are indeed well distinguishable in most of the two-dimensional space. Conversely, a not-negligible overlap among the samples of the two different classes is present near the graph origin, where all the states of muscle relaxation lies. As a

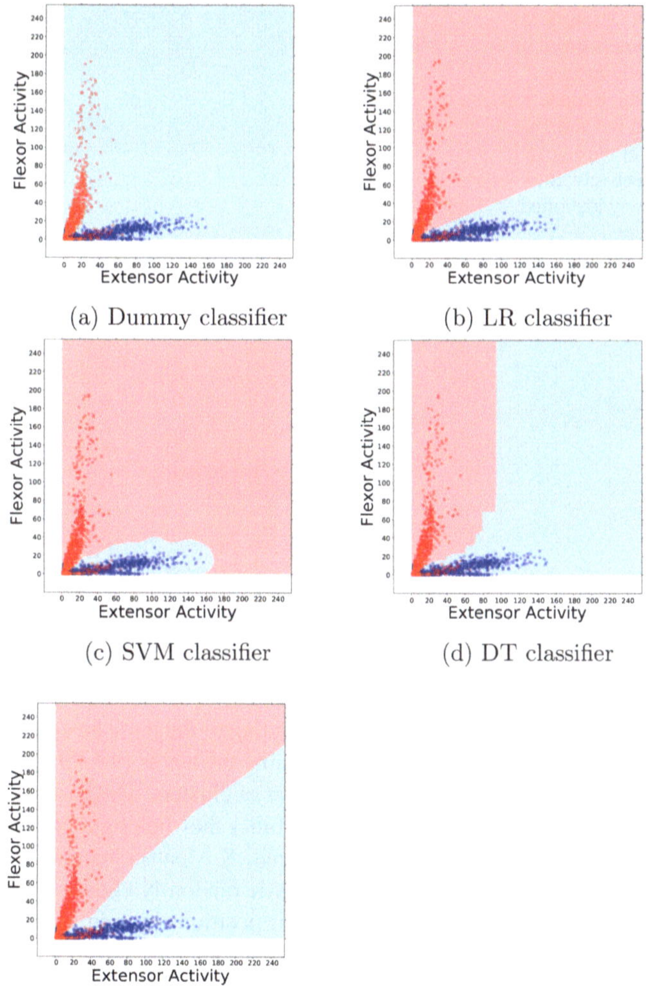

(a) Dummy classifier (b) LR classifier (c) SVM classifier (d) DT classifier (e) KNN classifier

Fig. 3.13 The decision boundaries from the testing of the five 2-class classifiers. In blue and red the opening and the closing intention respectively

result, it was almost impossible to distinguish one movement from the other during the initial moments of the muscle contraction.

Nevertheless, this is the state in which both movements always begin. Therefore, it was necessary to delimit this area by imposing a "resting" label on all the samples that fell into it. Since the rest state represented the system's safe mode—it does not imply any HES motion—each of such ambiguous points has been associated with the resting class to mitigate possible damages resulting from incorrect classifications.

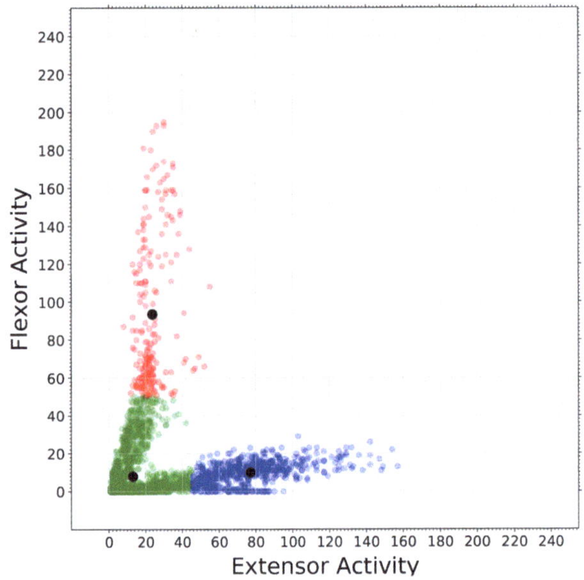

Fig. 3.14 The graph shows the clustering performed by exploiting the K-means technique as the K value was set to three. In red and blue, the intention of closing and opening respectively; in green, the newly identified resting intention. Furthermore, the black dots represent the optimized centroids for each cluster

The K-means clustering technique has been then used to confirm what had been qualitatively perceived from the decision boundaries of the four tested classifiers: the presence of three distinct classes corresponding to the three gestures to be recognized. Widely exploited in unsupervised learning problems [27–29], this technique partitions a not labeled dataset into a K number of clusters. The base idea is that the closer the data points are, the higher the possibility they belong to the same class is. Chosen K as the number of classes to be found, K-Means clusters the data points using a recursive algorithm. It, firstly, assigns K randomly chosen centroids to the different classes and, secondly, optimizes their positions by minimizing the sum of squared distances from every point of the cluster. Such an index can be seen as an indication of the dispersion of the samples assigned to a specific cluster. The "Elbow method" [30, 31] has been utilized to determine the optimum number of classes to be searched within the given dataset, which has turned out to be exactly three. Once the number of classes has been established, the algorithm has been executed on the original dataset to identify the different clusters; the results are visible in Fig. 3.14.

In light of the considerations mentioned above, the classification tuning has been performed again over the same algorithms. However, this time, the different classifiers have been trained and tested on a modified dataset consisting of the three classes resulting from the clustering. The results of this new attempt are reported in Fig. 3.15 and Fig. 3.16.

The resulting metrics report values that, on the whole, have improved up to 10% compared to those of the classifiers previously evaluated on the recognition of only two intentions: except for the Dummy classifier, all the other algorithms have metrics ranging from 92% to 94%. The apparent performance improvement is mainly shown

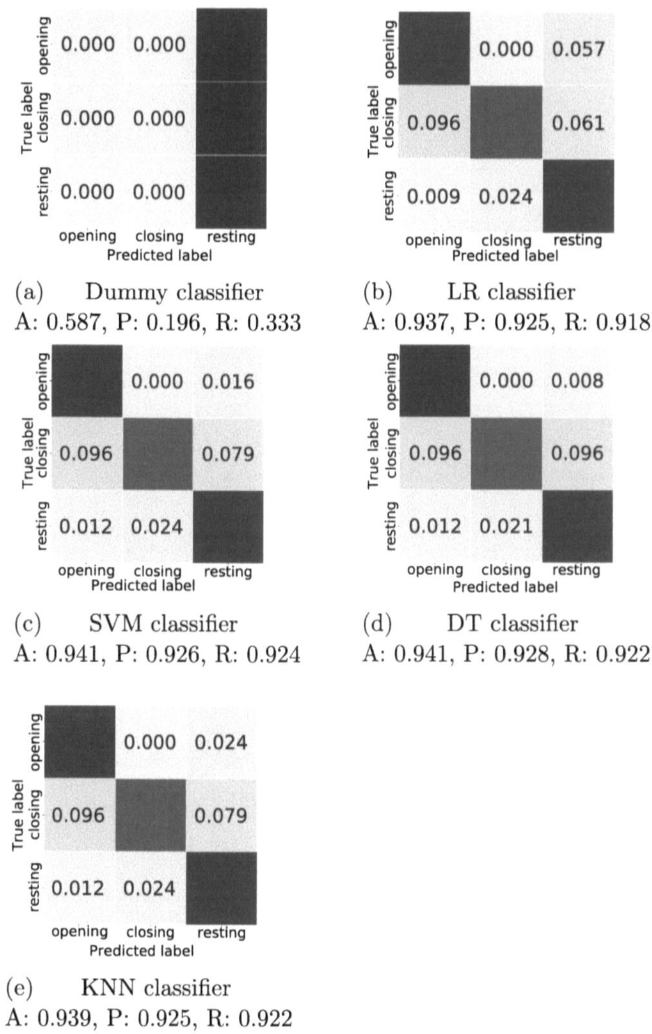

Fig. 3.15 The confusion matrices from the testing of the five 3-class classifiers. A stands for Accuracy, P for Precision, R for Recall

by the decrease in the number of false positives and negatives, represented by all the terms that are not on the main diagonal of the confusion matrices. This reduction is a direct consequence of adding the new resting class to the dataset, and the plots of the new decision boundaries confirm it. If these new plots are compared to the 2-class classifiers, no distinct differences in the level of class separation can be spotted where the sEMG signals are high. Conversely, remarkable changes in the class separation can be easily noticed right where the new points labeled as "resting" have been added. Thus acting, the number of harmful misclassifications has been reduced at

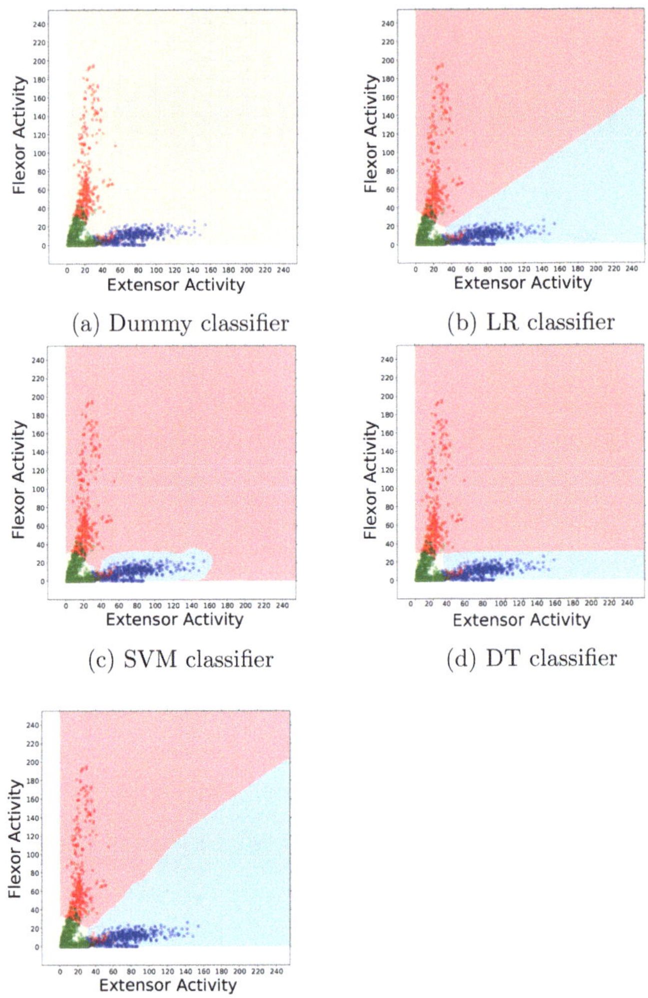

(a) Dummy classifier

(b) LR classifier

(c) SVM classifier

(d) DT classifier

(e) KNN classifier

Fig. 3.16 The decision boundaries from the testing of the five 3-class classifiers. In blue, red and green the opening, the closing and the resting intention respectively

the expense of a worse sensitivity to small muscle contractions. As a matter of fact, before the myoelectric activity can be classified into an actual motor intention, i.e., opening or closing, the muscle contraction has to exceed the area of the Cartesian plane close to the origin that contains all the resting-labeled samples.

Although the classification assumes so a sufficiently reliable functioning to undergo real experimentation, intrinsically, the management and rejection of misclassifications due to peaks of the signal caused by environmental disturbances remain unchanged. Indeed, through the use of the standard classifiers presented so far, the problem of upper delimiting the classes to avoid misclassifications from the recordings of possible artifacts has not yet been solved—the only case in which a complete class delimitation is shown in Fig. 3.16c but just for the opening intention. A particular two-dimensional algorithm has been found by rummaging the state of the art: it addresses classification problems, includes all the sought characteristics (i.e., that it shall not only separate at least three different classes in a two-dimensional plane but also circumscribe them in closed and limited areas), and recalls the constraints imposed by the clinical scenario (i.e., the classification shall be done in real-time while exploiting little computational power). the algorithm, known as "Point-in-Polygon", has been identified and adopted as the most viable path among the others present in the scientific literature.

Point-in-Polygon algorithm has been known since the early 1960s [32]. Even though it is widely used in fields such as computational geometry and computer graphics, it had never found application in the field of Wearable Robotics before. Turning to the functioning strategy, it checks whether a point in the plane lies inside or outside a given polygonal area. Two major versions of the base algorithm have been developed for this purpose by following two distinct approaches.

Ray Casting Version—Also known as the "Crossing number algorithm" or the "Even-odd rule algorithm" due to its operation logic, it is based on the Jordan curve theorem. First proposed in 1887 by the French mathematician Camille Jordan, it states that: "*any continuous simple closed curve in the plane, separates the plane into two disjoint regions, the inside and the outside*[3]" [33]. A direct consequence of this theorem is the mathematical certainty that any path from a point in one region to a point in the other region shall pass through the curve; it is on this observation that the algorithm's logic is built. Given a geometrical plane, a polygon, and a point on such a plane, be the point chosen at will inside or outside the polygon, by drawing from that point a straight line in any direction, the perimeter of the polygon will be intersected a certain number of times. By moving the point on this line, for each of these times it crosses the perimeter, it is known that it will pass from one region of the plane to the other. Hence, it is straightforward to notice that the point's initial region of belonging can be traced back from the number of intersections with the polygon's perimeter (see Fig. 3.17).

In practice, for simplicity, a ray casting to the right is performed, which means that, given the coordinates of the starting point, the ray is plotted, keeping its ordinate constant. The point belongs to the outer region if the ray crosses the polygon edges an even number of times; vice versa for an odd number of crossings. However, this technique presents some singularities—e.g., when one of the polygon's edges lays on the ray or when the ray passes through a vertex. As shown in Fig. 3.17, in the first case (e), the number of crossings is infinite, and the algorithm cannot assign a belonging region to the point; in the second case (a and c), there might be ambiguities in the

[3] Fun fact: even if it sounds pretty evident as a theorem, it has been very hard to verify. The proof Jordan gave in his famous textbook "Cours d'Analyze de l'École Polytechnique" turned out to be flawed, and the first valid proof was given almost 20 years later by the American mathematician Oswald Veblen, in 1905.

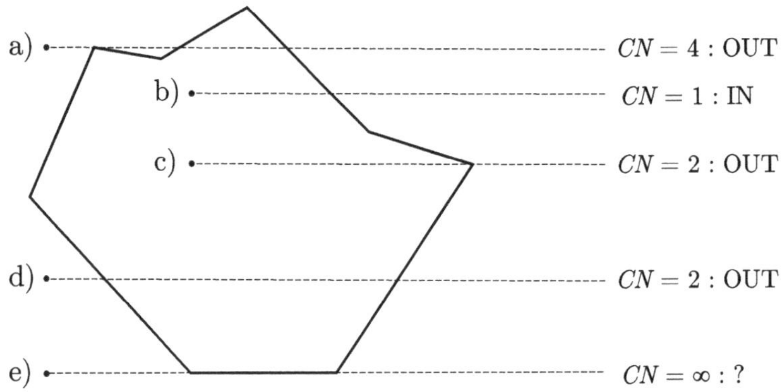

Fig. 3.17 The graphic representation of the operation principle of the ray casting version of the Point-in-Polygon algorithm. CN is the Crossing Number, i.e., the number of times the ray crosses the polygon perimeter: if it is odd, the point lies inside the polygon; the point lies outside if it is even. Case **a, b** and **d** are correctly classified; **c** and **e** are particular failure cases of this method

result since the intersection with a vertex is counted as a double crossing. Besides, when implemented on a computer, the result might be wrong even if the point lies in a small enough area around the perimeter.

Algorithm 3.2 Point-in-Polygon algorithm (Ray casting version)

Input:
N_v = number of the polygon vertices
$V_x[\ ], V_y[\ ]$ = arrays of coordinate of the vertices
T_x, T_y = coordinate of the test point

```
int j = N_v − 1
bool inside = false
for ( int i = 0; i < N_v; i ++) do
    if ( ( ( V_y[ ][i] < T_y and V_y[ ][j] ≥ T_y ) or ( V_y[ ][j] < T_y && V_y[ ][i] ≥ T_y ) ) and ( V_x[ ][i] ≤ T_x or V_x[ ][j] ≤ T_x ) ) then
        inside ^= ( V_x[ ][i] + ( T_y − V_y[ ][i] ) / ( V_y[ ][j] − V_y[ ][i] ) ∗ ( V_x[ ][j] − V_x[ ][i] ) < T_x )
    end
    j = i
end
return inside
```

The most common implementation of such an algorithm—presented in Algorithm 3.2—recursively checks for intersections with all the polygon's edges in turn. The main drawback of this method is that the result might be ambiguous when the ray passes precisely through one of the polygon's vertices. In this case, the ray will intersect two edges at their endpoints: the resulting number of the crossing will be even, and the point will be counted as external. While the result is correct for some of the cases (e.g., case a in Fig. 3.17), others require that the number of inter-

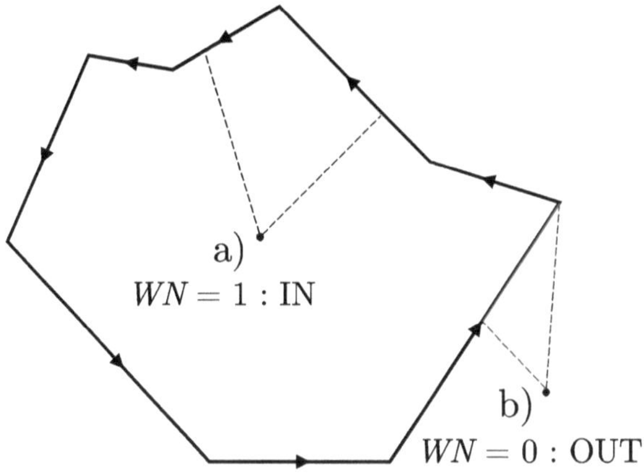

Fig. 3.18 The graphic representation of the operation principle of the winding number version of the Point-in-Polygon algorithm. *WN* is indeed the Winding Number, i.e., the number of complete counterclockwise turns that the segment (dashed in the figure), which has as its extremes the point to be tested and any point of the polygon perimeter, makes around the point itself along the entire perimeter: if this number is different from zero, the point lies inside the polygon; conversely the point lies outside if it is zero

sections is not counted more than once(as in case c in Fig. 3.17). A similar problem arises when checking for intersections with horizontal segments (case e). This is not generally a big issue since the computational speed provided by the Point-in-Polygon algorithm brings much more value than the one lost in lack of accuracy in these very particular cases. Nonetheless, the ambiguity can be easily solved by counting the intersection on a vertex only if the second vertex of the same edge lays below the ray.

Overall, the Ray casting version is a lightweight and high-performance algorithm for two-dimensional classifications. Its level of complexity and calculation time are directly proportional to the number of polygon vertices. Its robustness strongly depends on the finite arithmetics of the machine.

Winding number version—The second version of the Point-in-Polygon algorithm still lays the foundation of its operation logic on Jordan's Theorem. It is based on calculating the Winding Number (WN) of the given point for the given polygon. In Mathematics, this number is defined as the total number of times a closed curve in the plane travels counterclockwise around a certain point. In Automation, it forms the basis of the Nyquist stability criterion for unstable open-loop systems with a possible single or double pole in origin. WN is a characteristic of the curve-point pair rather than the individual geometric entities. According to this method, the point is inside the curve if the resulting Winding Number is different from zero. For this reason, this version of the algorithm is also called "Nonzero-rule algorithm". Figure 3.18 shows its base functioning.

Two are the main ways to compute the winding number. The first technique involves summing up the angles subtended by each side of the polygon [34]. It is generally slower than the ray casting algorithm since it involves inverse trigonometric functions, which are computationally heavy. Nonetheless, since the sum of all such angles in a polygon can add up to multiples of 2π only, the algorithm can be sped up by just keeping track of the number of quadrants that the polygon winds as it turns around the test point [35]. This second method, a direct improvement of the previous one, makes the speeds of the two versions of the Point-in-Polygon algorithm comparable. Besides, the winding number version provides the correct answer for non-simple polygons or particular cases, whereas the ray casting algorithm, as it has been presented, fails. Nevertheless, a more recent review of the ray casting [36] method has also smoothed out this difference, making the two versions practically equivalent.

Since there are no substantial differences in the execution speed or the results of the two codes, it has been arbitrarily chosen to use the Point-in-Polygon algorithm in its ray casting version.

The use of this type of classifier results in being particularly interesting. Indeed, it is possible to enclose only the opening and closing classes within well-defined areas—as shown in Fig. 3.19b—and implicitly assign the resting label to the remaining of the two-dimensional space. This way, the classification of the motor intentions becomes more selective and, consequently, more reliable in terms of the exoskeletal output. At the same time, the rejection of disturbances and unwanted movements [7] is automatically improved and has no additional computational costs. While maintaining performance comparable to the other classifiers—see Fig. 3.19a—the Point-in-Polygon also offers this additional safety level: all spurious electromyographic patterns, which do not selectively fall within the polygonal areas traced to delimit the intentions of opening and closing, are cataloged as resting and will not produce undesired effects on the HES.

However, one of the disadvantages of using this algorithm is the difficulty of efficiently automating the tracing of the polygons. Although the definition of polygonal decision boundaries for three classes in a two-dimensional plane represents a simple task for human intelligence, it is still challenging to accomplish with modern AI algorithms. One possible solution could be to develop a Deep Neural Network (DNN)—an Artificial Neural Network (ANN) prepared applying Deep Learning techniques—that can trace the polygons optimizing the separation of the classes. This would require extended time to train the network. Besides, it would not adapt very much to the clinical environment, characterized by a great diversity of patients. As a matter of fact, such diversity increases even more if the different types of pathologies and the different stages of the single disease are considered. For these reasons, the network would require frequent training on high-performance workstations, usually not available in clinical facilities, to match the addressed scenario. Furthermore, these steps require experience in Computer Sciences to be carried out in the best possible way. Instead, they will probably be performed by personnel with non-technical expertise, perhaps by the physiotherapists themselves. In light of this perspective, once abandoned the idea of automating the tracing of polygons, a custom GUI has been developed to provide a user-friendly tool for quick and easy data collection and tuning of the classification parameters. The above GUI runs on any

Fig. 3.19 The results of the testing of the Point-in-Polygon classifier

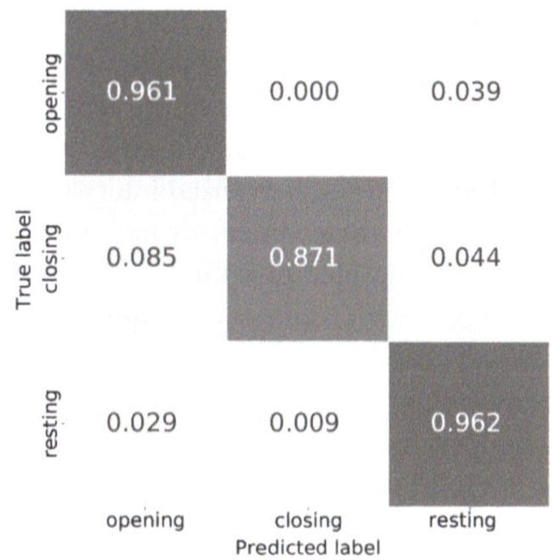

(a) Confusion matrix of the Point-in-Polygon classifier. A: 0.944, P: 0.933, R: 0.931.

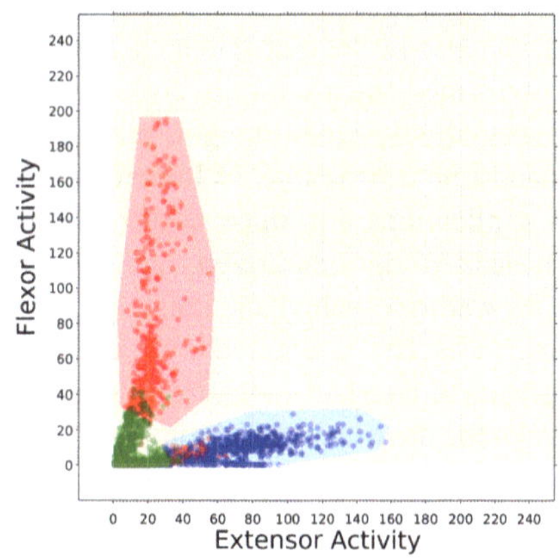

(b) Decision boundaries of the Point-in-Polygon classifier. In blue, red and green the opening, the closing and the resting intention respectively.

desktop PC or laptop and is designed to help clinical staff carry out the classifier training phase (polygon tracing) and upload the tuned model to the microcontroller board. The following section will present how it works in detail.

3.3.2 Graphical User Interface

The research for the classifier model that best suited the scenario under consideration has concluded with the adoption of the Point-in-Polygon algorithm in its ray casting version. Taking as inputs the number and coordinates of a polygon vertices and the coordinates of a given test point, it counts the total number of times a ray cast from the point crosses the polygon perimeter. Finally, it determines whether or not the point falls within the figure based on the Crossing Number. As previously mentioned, the choice of this classifier has forced its training and parameter tuning to be performed through external intervention. Since the HES has been designed for use in a clinical environment, it has been mandatory to develop a dedicated software application, namely a GUI, to help non-technicians personnel: (i) efficiently collect proper sEMG datasets; (ii) quickly identify the classifier parameters (i.e., the number and coordinates of the polygon vertices); (iii) straightforwardly program the microcontroller.[4]

Developed for the first time at the Xerox Palo Alto research laboratory in the late 1970s, a GUI is an interface through which users interact with electronic devices such as computers, handheld devices, or processors. This interface uses interactive components such as icons, menus, and other graphical objects and visual indicators to display information and related user controls. GUI representations are manipulated by a pointing device such as a mouse, trackball, stylus, or a finger on a touch screen. Still, it can also be navigated using a keyboard. Objects that resembled the operation to be performed and are intuitively recognized by users are depicted on the monitor using specially designed images, pictures, shapes, and color combinations. Unlike command-line instructions or Integrated Development Environments (IDEs), GUI-based applications are much easier to learn and use, mainly because commands do not need to be memorized, and users do not need to have any programming skills. Nowadays, such visual language has become commonplace in both operating systems and software applications. There are even IDEs—which are nothing more than graphical interfaces—entirely dedicated to developing custom GUIs.

One of the most used development environments for creating graphical interfaces and cross-platform applications is Qt Creator IDE. In the framework of this research activity, it has been chosen for its compatibility with all major desktop platforms and most mobile or embedded platforms. Qt Creator is a multi-platform IDE specifically designed for application development based on the C++ language. Released for the first time in 2009, over the years, it has evolved to become the hub of application development, integrating project management, source editing, graphics interface design, and third-party debugging and versioning tools. Qt Creator IDE is distributed by the Qt Company in two versions: one with an open LGPL license that involves some restrictions on distribution methods and one with a commercial license that also guarantees technical support for developers. The graphical interface for tuning the classifier has been designed using QT Creator IDE non-commercial distribution and targeting Microsoft Windows Operative System (OS) since it is the first desktop OS for dissemination.[5]

---

[4] The bracelet Arduino Nano board is the one in charge of classifying the sEMG signals and, therefore, the tuned classifier parameters shall be uploaded to it from the PC where the training phase is performed.

[5] https://gs.statcounter.com/os-market-share/desktop/worldwide (April 2021).

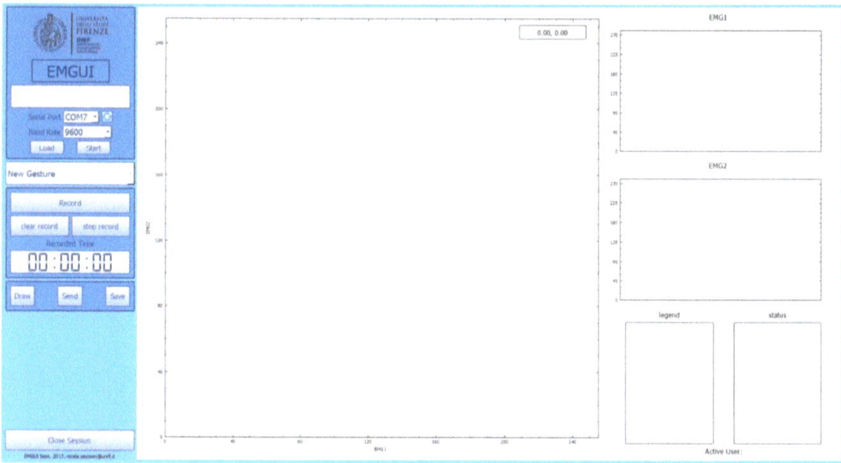

Fig. 3.20 The main window of the GUI developed for manually training the classifier. The interface is developed almost entirely on this unique view and the interaction takes place via keyboard and mouse pointer

Figure 3.20 shows the main window of the GUI. The four main sections of the interface shown in the figure are: (i) on the top left corner, the University logo and the application name; (ii) below that, the command layout that contains all the buttons through which commands are sent to the interface; (iii) in the center, an interactive two-dimensional Cartesian plane; (iv) on the right, from the top to bottom, two real-time graphics, the legend, and the status box.

As regards the command layout, starting from the top, there is a line of editable text to enter the title of the training session to be opened (e.g., *patient_name_date*). Right below, the commands for the initialization of the serial port: the interface has a button for automatic search of all the serial devices connected to the platform, the port is then selected from a drop-down menu. The interface memorizes the parameters of the last paired device and suggests them as the default choice during the opening of the further session. Further below, the interface initialization box ends with two buttons. "Load" opens the standard Windows folder browser allowing for selecting an old session and loads pre-recorded data from it. "Start"uses all the data previously entered to activate the other interface areas (disabled until "Start" is pressed) and initializes the serial communication. The serial connection can be either wired or Bluetooth. In the second case, the initialization will take a slightly longer time to complete because of the pairing.

Once "Start" has been pressed and the session begun, the data acquisition options can be managed with the tools provided under the initialization box.

If the session has started for a new user—hence without using the "Load" button—the central graph and the legend will remain empty. Conversely, the two graphs in the top right corner will start streaming the real-time envelope of the two sEMG signals and the "status" box will confirm the connection with the serial device (see

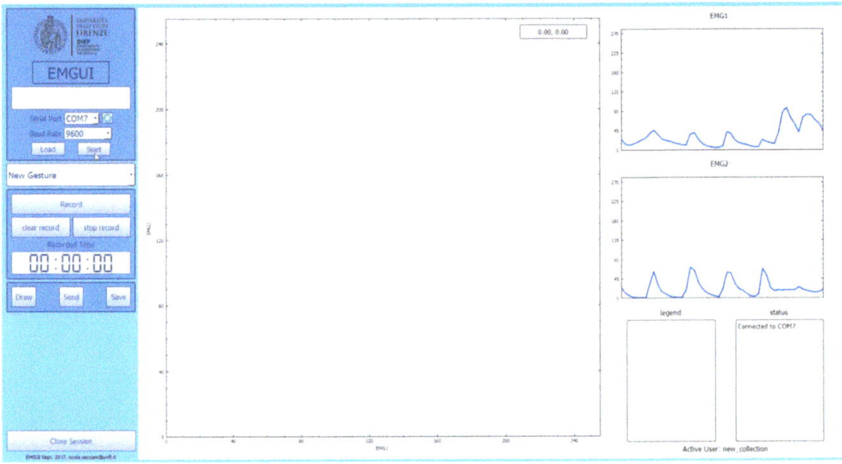

Fig. 3.21 The figure shows what happens on the interface once the "Start" key is pressed. In the top right corner the two graphs show the real-time trend of the electromyographic data collected by the two MyoWare^{TM} Muscle Sensors; in the bottom right corner, within the status box, the serial port and its connection status are shown; below the title of the current session

Fig. 3.21). Any connection errors will, in turn, be shown in red in the status box, and a new session shall be open to reset them. Suppose the session is instead opened for an already registered user. In that case, the behavior of the interface will be the same. Still, the user will see the previous acquisitions on the central graph.

The text line, reporting by default the "New Gesture" label, is a drop-down menu that contains all the gestures acquired up to that moment, whether they come from the current or a previous session. The menu can be explored by clicking on the small arrow on the panel right and scrolled with the mouse wheel. If already present in the list, it is possible to select a specific gesture and proceed with the acquisition of new data by pressing "Record"; this option overwrites the previously acquired dataset for that particular gesture. If the user does not desire to delete old data or no gestures have already been recorded, the new acquisition can be started while leaving the default label on the gesture drop-down menu. In this case, the user is asked to enter the gesture name before the acquisition starts. Choosing a name already within the gesture list will create a copy and will not overwrite the previous data. There are no limits to the number of gestures that can be acquired using the GUI. However, the constraint of being able to train the classification of only three hand gestures (opening, closing, and resting) remains. Instead, the number of excess gestures can be used to perform preliminary tests and evaluate the quality of the acquisition.

As reported in Fig. 3.22, once in the data recording mode, the timer will show the recording time while the central graph will be populated with electromyographic data from the two real-time plots. The myoelectric signals are shown in the graph as points whose coordinates are the values of the envelopes recorded by the two different sEMG sensors. Why the axes of the graph are limited to the range between 0 and 255 will be examined later on. The muscle activation of the *flexor digitorum* is

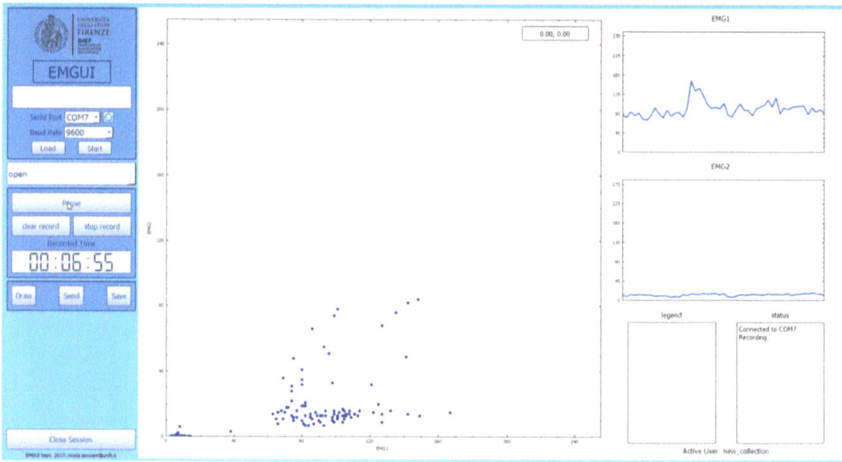

Fig. 3.22 The custom GUI in recording mode. On the left, the label of the gesture can be read in the drop-menu, just below, the "Pause" button has replaced the "Record" one and, below again, the timer that marks the time from the beginning of the acquisition can be seen. In the graph in the middle, the points show in real-time the muscular state of the two muscle bands under examination. Finally, within the status box, on the bottom right corner, a new line says "Recording...", confirming the current modality

reported on the ordinate axis. The abscissa instead shows the one from the *extensor digitorum*. Different gestures will be highlighted in different colors. As the recording proceeds, it is possible to pause it by pressing the "Record" button, which will now show the "Pause" label. The collection will then resume by pressing the same button once again. The "Clear record" button allows for deleting all the acquired data from the current gesture without interrupting the ongoing collection. The "Stop record" button has a double effect instead, depending on whether the current record contains data or not. If data is present, the button will interrupt the acquisition, save it in a text file (.txt) within a specific folder, and, finally, add the name of the gesture in the list of the acquired ones. Inside the text file the sEMG envelopes are saved trivially as pairs of integers separated by a horizontal tab; samples belonging to different sample times are reported on different rows. On the contrary, when there is no data recorded (e.g., right after it has just been cleaned up with the "Clear record" button), the stop button ends the collection and deletes all its tracks. This process can also be used to delete old recordings that are no longer useful. The points of the gestures correctly saved will stay on display on the central Cartesian plane until the end of the session, or until they are deleted using the method described above. The status box will show the history of the actions carried out during the acquisition.

Once the acquisition phase of the (training) dataset has been completed, the further step involves the manual drawing of two polygonal areas around the points that identify the closing and the opening gesture, as shown in Fig. 3.23. To do so, the two gestures (one at a time) shall be selected from the drop-down menu and the"Draw"

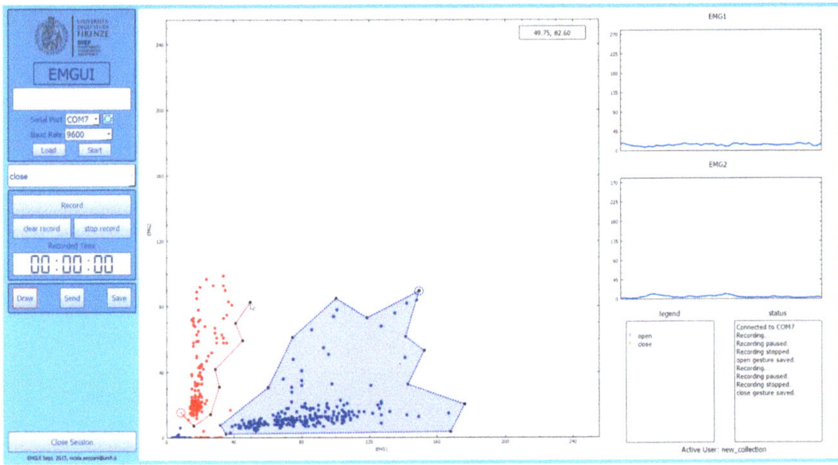

Fig. 3.23 The drawing mode that allows to trace the polygons on the central graph. Vertices are drawn by clicking on the graph, polygon sides are previewed while moving the mouse and defined when drawing the second vertex, and, finally, on the top right corner of the Cartesian plane the coordinates are shown

button pressed, unlocking the interaction with the central graph to draw the relative polygon. A red border around the button will highlight the entering of the drawing mode. The first vertex to be drawn will be circled in the same color as the gesture points. The circle helps close the polygon: it is sufficient to insert any other vertex within this circumference to end the drawing and close the polygon on the first drawn vertex. The drawing mode will be exited at this point, the area inside the polygon will be colored with a light shade of the gesture color, and the "*gesture_name gesture saved.*" line will be appended to the status box. The same steps can be repeated to draw the other polygon by selecting a different gesture from the list. It is also possible to delete an already drawn polygon by pressing the "Draw" button once selected the corresponding gesture.

Next to the "Draw" button, the keys labeled "Send" and "Save" are, respectively, in charge of uploading the classifier parameters to the microcontroller board and saving the the polygon vertices on a text file.

During regular use of the interface, the last button to press is labeled "Closed Session" in the bottom left corner. Doing this proceeds with closing the session, clearing the serial buffer, and returning the interface to the original state shown in Fig. 3.20.

The data transmission protocol that manages the communication between the sEMG bracelet and the GUI will now be described. Why the axes of the central graph are confined between 0 and 255 will also be explained.

The first thing to notice is that saving data directly on the microcontroller is mandatory so that they are automatically reloaded from the internal memory without having to connect again to the graphical interface. However, data needs to be saved in a specific format on the Arduino Nano board. The only non-volatile storage that Arduino Nano can offer is called EEPROM memory, which stands for Electrically Erasable Programmable Read-Only Memory.[6] Contrary to what the acronym suggests, EEPROM memory can be programmed byte to byte, exploiting particular programming signals. Generally, this process is facilitated by medium-high level functions; the Arduino IDE provides a ready-to-use library for such a task. Unlike flash memory, EEPROM memory cannot handle multi-byte operations but is instead designed to manage single-byte operations. Nonetheless, the number of life cycles can reach up to one million; in contrast to flash memory which stops at tens of thousands. While this ensures a long duration of its integrity, it also makes storing information more challenging. In light of these reasons, it has been decided to map the 10-bit sEMG signal returned by Arduino Nano's ADC to 8 bits before plotting them on the Cartesian plane. At the same time, the graph has been designed to host points with coordinates limited between 0 and 255. The classifier parameters, i.e., the polygons' vertices, will be hence already in the correct format to be appropriately stored in the microcontroller's EEPROM memory.

Uploading such parameters to the Arduino Nano is performed in individual packets of one byte, each in a precise sequence. The first outgoing value represents the number of polygons. At the moment, two is the only number possible to send; still, the protocol has been intentionally left open to upgrades. After the first byte, the number of the first polygon vertices is sent. Assuming this number is K, the following 2K values to be sent are the couples of the vertices' coordinates in the drawing order (abscissa first). The same is then repeated for the second polygon. All these values are transmitted interspersed with the separator character ";" to facilitate the microcontroller in receiving and parsing them. At the end of the data transmission, a line with the number of sent bytes is added to the status box to confirm the correct execution of the command Fig. 3.24.

Through the presented procedure, the training phase of the Point-in-Polygon classifier is performed manually using a custom-made GUI running on Microsoft Windows. The classifier's parameters are transferred from the PC to the microcontroller via serial communication over a Bluetooth bridge. They are then stored on the EEPROM memory. Classification accuracy and disturbances rejection are guaranteed, setting these parameters on the specific patient's needs. Since these necessities are very likely to vary over time thanks to therapy progress or due to the pathology evolution, it was essential to provide the clinical staff with an easy-to-use tool to quickly adapt them to different conditions. For these reasons, the classifier's tuning is meant to be done manually by clinical personnel who know the users' medical history; it is also likely to be performed before the patients get discharged.

Once the classifier has been prepared for the real-time action, the code on the microcontroller is changed by pressing the Arduino Nano board reset. This secondary algorithm no longer sends sEMG signals outwards but performs the classification instead. After collecting and mapping in the [0 255] interval the signals, it combines

---

[6] The ATMega328p microprocessor, the one at the core of the Arduino Nano board, offers 1 kB of EEPROM memory.

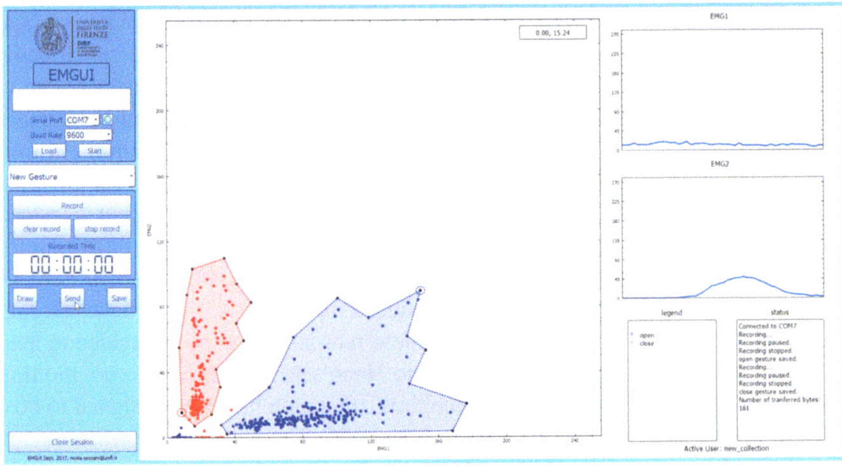

Fig. 3.24 A new line in the status box indicates the total number of bytes sent to the microcontroller pressing the "Send" button. Immediately after, the serial communication is interrupted to avoid disturbances that could corrupt the data. The classifier training session is hence concluded

them as the test point coordinates whose position in the plane has to be tested. The code loads the polygon vertices into its flash memory from the EEPROM memory the first time this code is turned on. Every sample time, the position of this point is evaluated based on whether or not it belongs to a specific area. Classification is then performed in an exclusive way for preventive purposes. As a matter of fact, although no constraints have been imposed on the reciprocal position of the two different polygons—which can also overlap —, each point shall belong to only one of them to be classified as an open or close intention. All points that simultaneously belong to both polygons or none of them are classified as rest intentions. The classification is translated into a numeric code—each class is assigned with an integer number—that is, finally, transmitted to the paired Bluetooth device as a single byte 50 Hz.

All this occurs on the bracelet microcontroller that can be interfaced with a desktop PC for monitoring purposes or directly to the controller module for actuating the exoskeleton. In the second case, the user can now control the exoskeleton motion by reproducing muscular signals as similar as possible to the ones used to train the classifier. Like learning how to drive a car or ride a bike, the correct, quick, and efficient use of the system takes time to be mastered and needs training not only for the classifier but also for the user.

A final training phase has been thought to facilitate the learning process before letting the patient actually wear and use the HES. Another GUI has been developed to ensure that this additional step would not have been too tedious. Instead of resorting to the Qt Creator IDE again, Processing 37 has been used. It supports both 2D and 3D graphics. It is provided with OpenGL integration for accelerated 2D and 3D and is well documented. Besides, it can be readapted to different platforms (GNU/Linux, Mac OS X, Windows, Android, ARM). A simple 3D hand model is displayed in this graphical interface. The mouse can be used to interact with it and dynamically change the camera view. The hand motion is directly correlated to the type of user's intention classified by the electromyographic bracelet. As it would happen when controlling the exoskeleton, it is possible to control the complete movements and all the intermediate positions. The speed of the hand motion can be adjusted according to the user's taste, and it is then translated into the actual speed of the exoskeleton movements.

References

1. Merletti R, Parker PA, Parker PJ (2004) Electromyography: physiology, engineering, and non-invasive applications, vol 11. Wiley
2. Mills K (2005) The basics of electromyography. J Neurol Neurosurg Psych 76(2):1132–1135
3. De Luca CJ (1979) Physiology and mathematics of myoelectric signals. IEEE Trans Biomed Eng 6:313–325
4. Garcia MC, Vieira T (2011) Surface electromyography: Why, when and how to use it. Revista andaluza de medicina del deporte 4(1):17–28
5. Stegeman D, Hermens H (2007) Standards for surface electromyography: The european project surface emg for non-invasive assessment of muscles (seniam). Roessingh research and development, Enschede, pp 108–12
6. Li G, Li Y, Yu L, Geng Y (2011) Conditioning and sampling issues of emg signals in motion recognition of multifunctional myoelectric prostheses. Ann Biomed Eng 39(6):1779–1787
7. Li X, Chen S, Zhang H, Samuel OW, Wang H, Fang P, Zhang X, Li G (2016) Towards reducing the impacts of unwanted movements on identification of motion intentions. J Electromyogr Kinesiol 28:90–98
8. Smith LH, Hargrove LJ, Lock BA, Kuiken TA (2010) Determining the optimal window length for pattern recognition-based myoelectric control: balancing the competing effects of classification error and controller delay. IEEE Trans Neural Syst Rehabil Eng 19(2):186–192
9. Phinyomark A, Khushaba RN, Ibánez-Marcelo E, Patania A, Scheme E, Petri G (2017) Navigating features: a topologically informed chart of electromyographic features space. J R Soc Interface 14(137):20170734
10. Phinyomark A, Khushaba RN, Scheme E (2018) Feature extraction and selection for myoelectric control based on wearable emg sensors. Sensors 18(5):1615
11. Spiewak C, Islam M, Zaman MA-U, Rahman MH et al (2018) A comprehensive study on emg feature extraction and classifiers. Open Access J Biomed Eng Biosci 1(1):17–26
12. Bishop CM (2006) Pattern recognition and machine learning. Springer
13. Kotsiantis SB, Zaharakis I, Pintelas P (2007) Supervised machine learning: a review of classification techniques. Emerg Artif Intell Appl Comput Eng 160:3–24
14. Hosmer DW Jr, Lemeshow S, Sturdivant RX (2013) Applied logistic regression. Wiley
15. Cortes C, Vapnik V (1995) Support-vector networks. Mach Learn 20(3):273–297
16. Smola AJ, Schölkopf B (1998) A tutorial on support vector regression. Stat Comput 14(3):199–222
17. Safavian SR, Landgrebe D (1991) A survey of decision tree classifier methodology. IEEE Trans Syst Man Cybern 21(3):660–674
18. Ho TK (1995) Random decision forests. In: Proceedings of 3rd international conference on document analysis and recognition, vol 1, pp 278–282
19. Cover T, Hart P (1967) Nearest neighbor pattern classification. IEEE Trans Inf Theory 13(1):21–27
20. Beyer K, Goldstein J, Ramakrishnan R, Shaft U (1999) When is "nearest neighbor" meaningful? In: International conference on database theory, pp 217–235

21. Secciani N, Bianchi M, Ridolfi A, Vannetti F, Allotta B (2019) Assessment of a hand exoskeleton control strategy based on user's intentions classification starting from surface emg signals. In: international symposium on wearable robotics, pp 440–444
22. Secciani N, Bianchi M, Meli E, Volpe Y, Ridolfi A (2019) A novel application of a surface electromyography-based control strategy for a hand exoskeleton system: A single-case study. Int J Adv Rob Syst 16(1):1729881419828197
23. Secciani N, Topini A, Ridolfi A, Meli E, Allotta B (2020) A novel point-in-polygon-based semg classifier for hand exoskeleton systems. IEEE Trans Neural Syst Rehabil Eng
24. Englehart K, Hudgins B et al (2003) A robust, real-time control scheme for multifunction myoelectric control. IEEE Trans Biomed Eng 50(7):848–854
25. Montagnani F, Controzzi M, Cipriani C (2016) Independent long fingers are not essential for a grasping hand. Sci Rep 6:35545
26. Pedregosa F, Varoquaux G, Gramfort A, Michel V, Thirion B, Grisel O, Blondel M, Prettenhofer P, Weiss R, Dubourg V et al (2012) Scikit-learn: machine learning in python. J Mach Learn Res 12:2825–2830
27. Forgy EW (1965) Cluster analysis of multivariate data: efficiency versus interpretability of classifications. Biometrics 21:768–769
28. MacQueen J et al (1967) Some methods for classification and analysis of multivariate observations. In: Proceedings of the fifth Berkeley symposium on mathematical statistics and probability, pp 281–297
29. Jain AK (2010) Data clustering: 50 years beyond k-means. Pattern Recogn Lett 31(8):651–666
30. Thorndike RL (1953) Who belongs in the family? Psychometrika 18(4):267–276
31. Kane A (2012) Determining the number of clusters for a k-means clustering algorithm. Indian J Comput Sci Eng 3(5):670–672
32. Shimrat M (1962) Algorithm 112: position of point relative to polygon. Commun ACM 5(8):434
33. Korniłowicz A (2007) A proof of the jordan curve theorem via the brouwer fixed point theorem. Mech Math Appl 6(1):33–40
34. Weiler K (1994) An incremental angle point in polygon test. Graphics gems IV 4:16
35. Hormann K, Agathos A (2001) The point in polygon problem for arbitrary polygons. Comput Geom 20(3):131–144
36. Galetzka M, Glauner P (2017) A simple and correct even-odd algorithm for the point-in-polygon problem for complex polygons. In: 12th international joint conference on computer vision, imaging and computer graphics theory and applications (VISIGRAPP 2017)

Chapter 4
Tests and Results

All the electronic changes presented in Sect. 3.2 have been followed by a profound mechanical redesign of the exoskeleton architecture and transmission system. This process has also addressed some of the problems that emerged while testing the first device (Sect. 2.3). Although they have not been the subject of this thesis work, they will be briefly described in this chapter for the sake of thoroughness. Afterward—in Sect. 4.1 and Sect. 4.2—the chapter will focus on the actual testing phase of the HES. The experimental protocol will be described in detail, and the solutions introduced to tackle the significant issues that have arisen during these new tests will be emphasized. At the end of each section, the test results will be presented in terms of device usability aiming at handling daily-use objects.

From the accurate study of the design process that has led to the production of the first HES prototype, the ergonomics and kinematics of the coupling between the hand and the exoskeleton have undergone significant improvements. Over the years, these improvements have resulted in an evolutionary process that has not only radically changed the appearance and equipment of the device [1, 2] but has also made it considerably easy to adapt the exoskeleton to the most diverse hand sizes, including those of children [3, 4]. Besides, the wise choice producing the various components by additive manufacturing in ABS has made the replacement of broken or old parts particularly cheap and quick.

The new prototype, shown in Fig. 4.1, is at first glance very different from the one shown in Fig. 2.4. Among all the differences, it is also possible to notice that the planar mechanisms of the fingers, although slimmer than the previous ones, keep the same kinematic structure. Each finger mechanism is composed of a planar single-DOF kinematic chain that acts on the middle phalanx to make the finger perform a complete flexion/extension movement, translated into an average ROM for the MCP rotation of about 80°. The device is still actuated through a cable-driven transmission.

N. Secciani, *sEMG-based Control Strategy for a Hand Exoskeleton System*, Springer Theses, https://doi.org/10.1007/978-3-030-90283-4_4

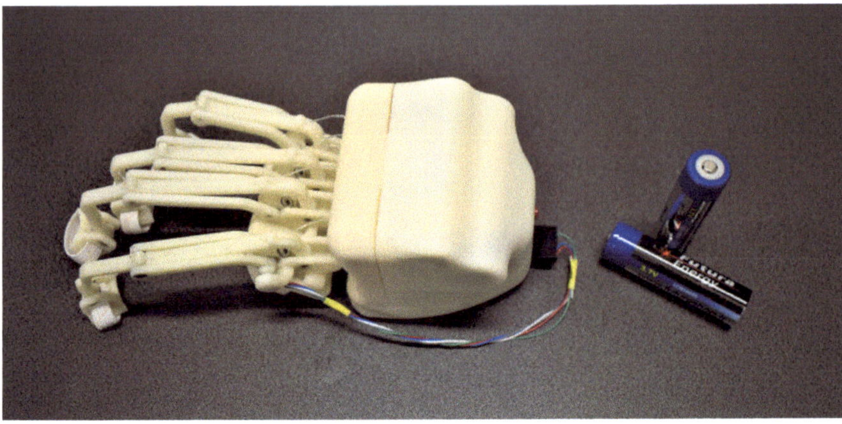

Fig. 4.1 The final version of the assistive HES developed at the UNIFI DIEF. On the right, the two lithium-ion batteries that power the device

The first notable difference is that this final version has been designed to have a modular structure. The motor and control box and the module that collects all the finger mechanisms are two separate and independent entities. The main idea behind this design choice is to minimize downtime for maintenance, whether it is programmed or not.

The mechanisms of the fingers are the part that will be mainly subject to replacement as the weaker part from a mechanical point of view and hence easy to break. Each finger mechanism has been designed to bear a maximum load of about 15 N applied to the contact point—a value considered an appropriate force output for manipulating everyday objects during normal activities [5]. Besides, since the patient's status may evolve in time, the trajectories of the fingers may change during the use of the device, and, therefore, the geometry of the finger mechanisms might need to be revised. Finally, the same actuation box can be used for different but similar hand sizes, thus streamlining both production and design processes.

As shown in Fig. 4.2, the motor box now incorporates all the electronics presented in Sect. 3.2 and the batteries, which are therefore no longer located along the forearm. Although this means a worse mass disposition, which is now concentrated on the hand—the miniaturization process is left to future investigations—it allows for the complete encapsulation of the power supply, making the system more compact and improving its wearability. The power supply stage, consisting of two rechargeable 3.7 V lithium-ion batteries connected in series with a capacity of 2600 mAh each, has been designed to last from one to two hours depending on the usage. Also, the actuation system has had to undergo changes that have drastically changed its functioning. Reducing the number of motors from four to one has led to the necessary redesign of motion transmission from the actuator to the finger mechanisms. In the new actuation system, the motor no longer acts directly on the cables. Instead, it sets in motion a secondary shaft through a toothed belt drive. Four pulleys of different

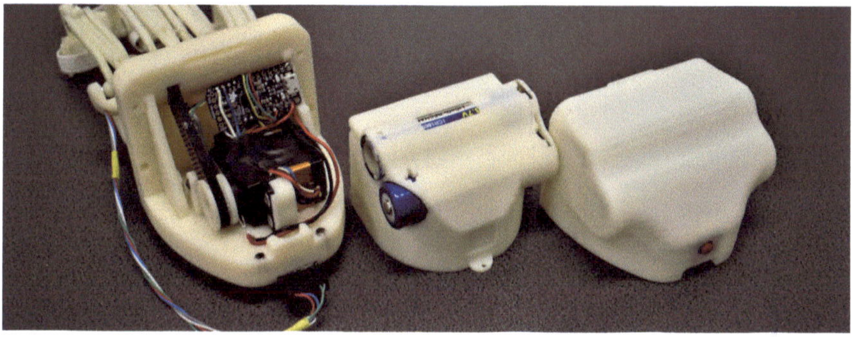

Fig. 4.2 An exploded view of the new motor box: on the left, the innermost module houses the actuator, the transmission system, and the control electronics; in the centre, the power stage contains the batteries and it is the only stage accessible to the user once the system has been delivered; on the right, the case enclosing the other modules, equipped with a switch button

diameters are mounted on this shaft and distributed along its length: rotating at the same angular speed, they wind and unwind different quantities of cable. This adjustment allows for synchronous motion of all the fingers, although they have different dimensions and, thus, their trajectories require different cable lengths.

Another meaningful change regards the additional passive DOF that has been introduced in joint ① in the form of a second revolute joint —please refer to Fig. 2.1 for joint and component nomenclature. In particular, this revolute joint has the axis lying on the flexion/extension plane of the finger. It has been added to passively align the mechanism to the finger and pander its natural movement of adduction/abduction passively. This addition has resulted not only in improved ergonomics and user's comfort but also allows for the alignment of the plane of action of the mechanism with the flexion/extension plane of the finger and helps them to be arranged more efficiently for the object grasping. This inclusion did not affect the total number of DOF of the finger mechanism and its kinematics, for it continues to be integral with the finger and to act in the same two-dimensional plane. As a result, the new HES has eight independent DOF: four actively controlled and four passive, uniformly distributed on all four long fingers (two per finger mechanism). The average ROM for the MCP rotation is about 80° on the flexion/extension plane. In comparison, it is roughly 15° on the adduction/abduction one.

The last significant variation from the original prototype concerns the ergonomics of the system dressing. As mentioned in Sect. 2.3, the previous device was placed in position, over the hand, using a sports glove to which it was sewn. The glove's elasticity, however, had a double effect: it ensured excellent safety of the coupling as it was intrinsically compliant with any possible displacement between the hand and the exoskeleton; conversely, this same characteristic made the exoskeleton produce different trajectories from those for which it was designed since each displacement modified the relative position between joint ① and the MCP joint indeed. In addition to being uncomfortable and limiting for the user—forced to unnatural movements—

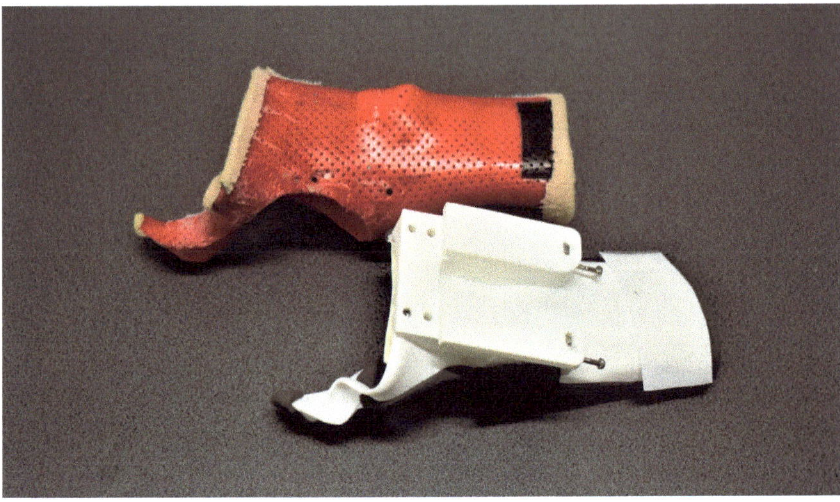

Fig. 4.3 On top, the anatomical wrist splint thermoformed on the patient's anatomy; below, the 3D printed reproduction in ABS incorporating a slide and threaded holes for mounting the two HES modules

this excessive mobility of the exoskeleton did not ensure reliable repeatability of the motion. A wrist splint has been used to provide a sufficient rigid interface base with the upper limb to overcome this problem.

A medical splint is a device that—when applied over or around an injured segment—increases, improves, or controls the impaired function. They are made by a trained therapist, using thermoplastic material that can be modeled even at relatively low temperatures (about 75°). Splints are modeled directly on the patient's anatomy and have significant advantages for the therapy as their wearing is a rehabilitation treatment in itself. Moreover, they are light, resistant, easily removable, easily washable, and can be continuously reviewed and modified according to the new requirements of the treatment plan. There are different types of splints: (i) the immobilization ones aim to maintain a correct posture of the limb to limit the onset of deformity and to protect the repair of injured tissues; (ii) the ones for protected mobilization are designed to selectively neutralize the proximal or distal joints to facilitate specific muscle or motor activity; (iii) the integrative ones facilitate a function by compensating with their rigidity the presence of paralyzed muscles or even prevent the appearance of deformities caused by the imbalance between the agonist and antagonist muscles; (iv) the ones for joint recovery promote, as the name suggests, joint recovery in cases of rigidity. The exploited splint (see Fig. 4.3) can be considered a combination of the typologies (ii) and (iii). In this case, the main task requested for the splint was to stiffen the wrist articulation to discharge the interaction forces between the exoskeleton and the hand throughout the forearm. It also concentrated the HES action on the fingers and gave greater stability to the system

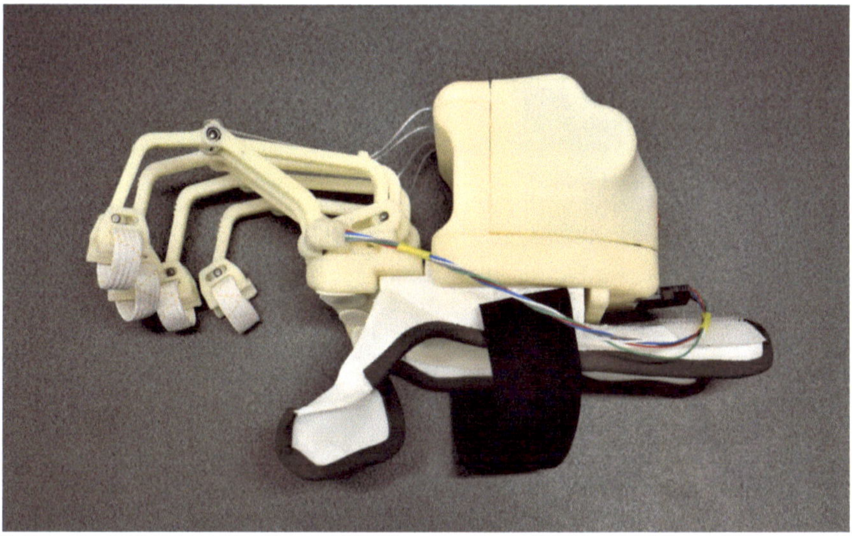

Fig. 4.4 The final HES mounted on the tailor-made splint

(protected mobilization). Besides, it provided rigid support to keep the thumb in a semi-opposition position to the palm facilitating object grasping and mitigating tendon retraction (integrative).

However, the splint was still not suitable as an anchor base for the exoskeleton, just as it came out of the thermoforming process. A three-dimensional CAD model has been created to allow for the modeling of the upper surface while maintaining the anatomical correctness of the lower surface. This process has taken place by exploiting a measuring arm equipped with a 3D laser scanner. Finally, the housing for the finger module and a magnetic slide for anchoring the motor box have been designed on the CAD model (both visible in Fig. 4.3), and a 3D-printed version of the splint has been produced. Figure 4.4 shows the complete version of the HES made within the MDM Lab at the UNIFI DIEF. The device—whose cost and weight have remained almost unchanged since the previous prototype (excluding the sEMG bracelet)—is fixed to the upper limb through two Velcro straps.

4.1 First Test Session

In the form of a pilot study, a first test session has been carried out at the Don Gnocchi IRCCS Rehabilitation Centre once the prototype's development was completed. The purpose of this study was to verify the actual impact of the exoskeleton use in object handling when worn by a real patient. Given the de facto exploratory nature of these tests, the test sample's minute size, and the low repetitiveness of the classifier

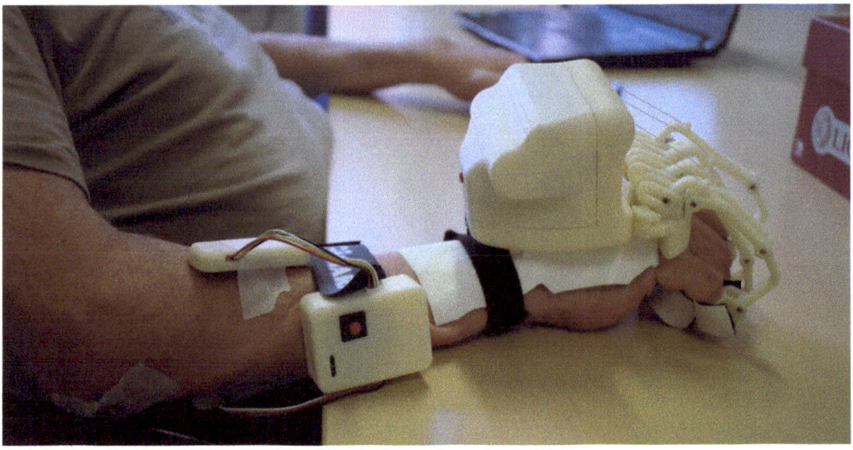

Fig. 4.5 The HES and the sEMG bracelet worn by the test subject

training, no statistical analysis on the classification accuracy has been performed. The test focused instead on manipulation tasks. One subject (male, aged 54, 1+ Modified Ashworth Scale [6]) has been enrolled for the study. The patient suffers from birth from SMA type II, which is a neuro-degenerative disease with several possible different outcomes. In this specific case, the pathology produced selective damage to the extensor muscles of both hands, causing a clenched fist deformity and resulting in hand opening impairment due to tendons' retraction. Before taking part in the tests, the subject has been provided with a written informed consent form and an information sheet.

Experimental setup—Aiming to provide a comfortable and high customized coupling system on the patient's limb, the ergonomics of the device has been improved by housing the HES on a 3D-printed splint, directly manufactured starting from the user's anatomy. Such a tailor-made interface has been fixed to the user's upper limb employing a Velcro fastening, as visible in Fig. 4.5, to assure a stable kinematic coupling between the exoskeleton and the hand. The sEMG bracelet has been previously worn by the user, and the two EMG sensors have been placed, with the help of a physiotherapist, on the *extensor digitorum* and *flexor digitorum* muscle bands exploiting adhesive Ag/AgCl electrodes. However, since the subject could not provide reliable signals concerning fingers' motion without feeling annoyance, the remarkable cross-talk component from the muscular activity of the wrist has been used instead. Before starting the test, the subject has conducted a complete training session of the classifier exploiting the GUI described in Sect. 3.3.2 under the supervision of the same physiotherapist[1] who assisted the dressing. Firstly, the sEMG signals related to the flexion and extension of the wrist have been recorded and labeled respectively

[1] The physiotherapist has been taught how to use the GUI to properly tune the classifier parameters in two practice sessions for a total of four hours the week before the tests.

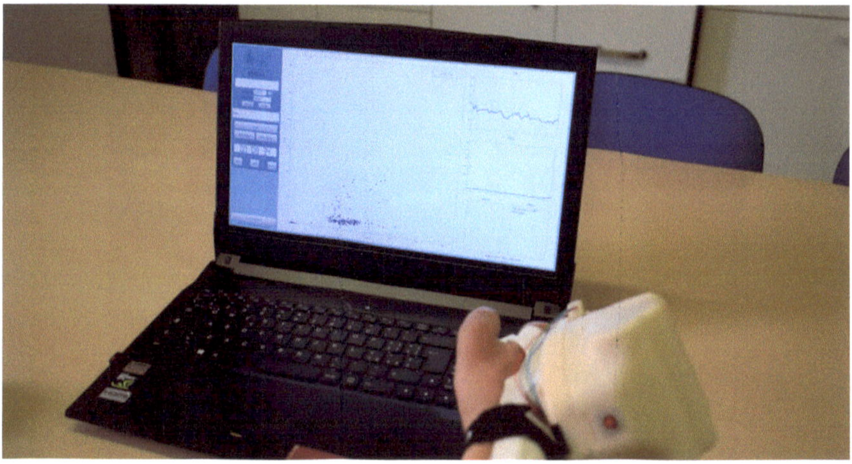

Fig. 4.6 The figure shows the collection of the sEMG signal during the training phase which has preceded the test session

as hand closing and hand opening intention (the process is shown in Fig. 4.6). This phase had necessarily to be performed while the subject was wearing the whole system since the dressing of the device itself influences the overall muscular activity. Secondly, the physiotherapist has drawn the gesture polygons in the way he thought best. Thirdly, the patient has virtually tested the control experience by commanding the motion of the 3D hand model. A user-training approach has been followed to teach the patient to consistently produce EMG patterns corresponding to clear commands for the exoskeleton. The transition to the actual test phase has taken place only when the subject expressed positive feedback on the reliability, reactivity, and sensitivity of the classification output; the training process was repeated until that moment.

Experimental protocol—The Single-Case Experimental Design methodology [7] has been chosen for the study: 10 objects of different sizes and shapes from those of daily use (i.e., an apple, a tangerine, a tennis ball, a mug, a 0.5 l bottle of water, a door handle, a spray can, a pot, a felt-it pen, a smartphone) have been selected for a chronometric functional task test. The subject was sitting in front of a table where the mentioned objects have been previously placed (Fig. 4.7). The subject has been asked to grasp each object and move it on a standard shoebox ($12.5 \times 28.5 \times 10$ cm) placed 10 cm away from the initial position. For each object, a total of five trials have been conducted, and the average grasping time was calculated. Figure 4.8 shows the subject while carrying out one of the chronometric tasks.

Results—The results of the first test session are reported in Fig. 4.9. As expected, the actual time to complete the tasks resulted in being relatively high compared to reasonable standards for able-bodies. However, it is more interesting to note how these results give an overview of the behavior of the HES as the type of the handled

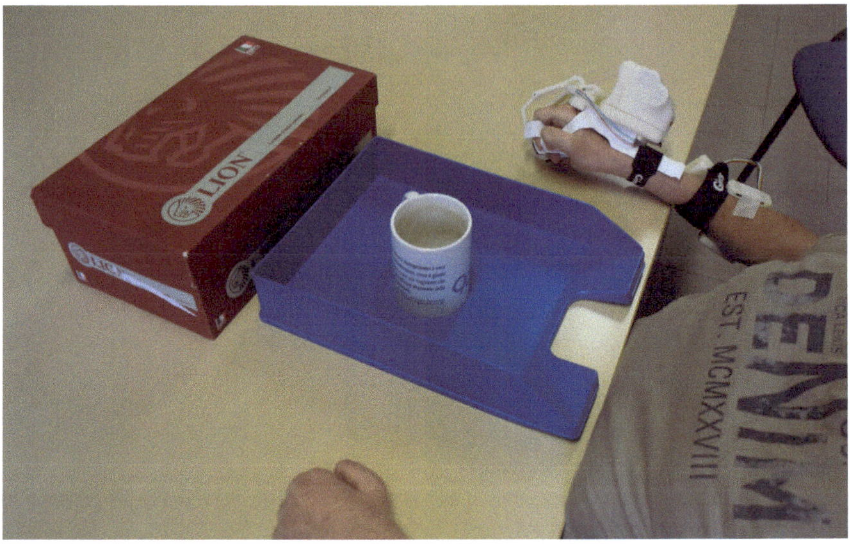

Fig. 4.7 An example of the test scenario: the subject is seating in front of a blue dais and a shoe box, in the center of the blue dais one of the test objects is placed; the subject shall move the test object over the shoe box only using the upper right limb assisted by the HES

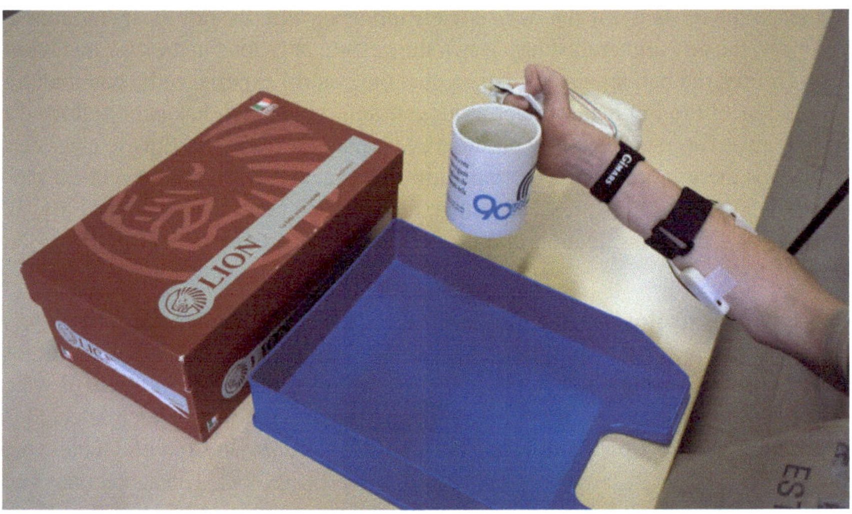

Fig. 4.8 The HES user as he moves an empty mug over the shoe box

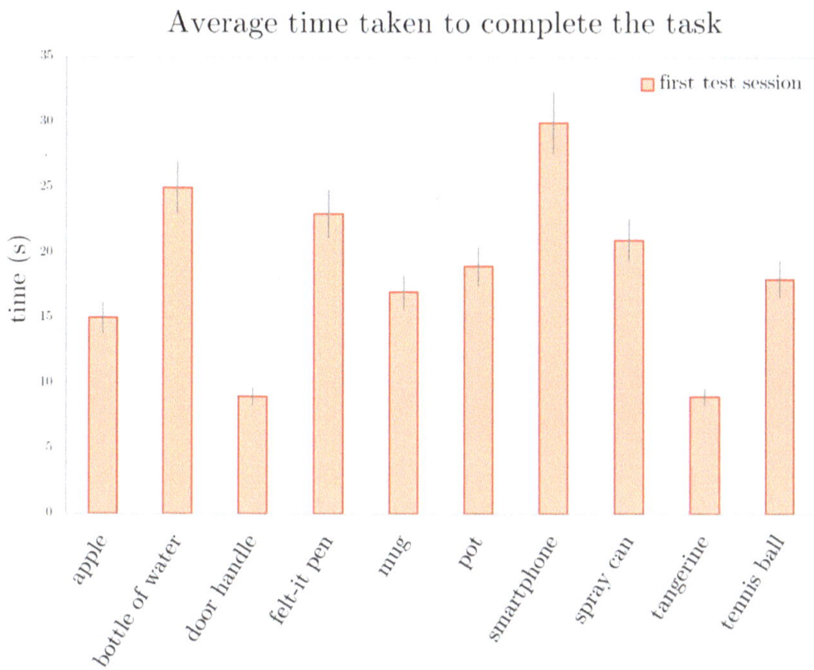

Fig. 4.9 The graph shows the results of the chronometric tasks test. The histogram reports on the x-axis the objects involved and on the y-axis the average time (rounded to the second) taken to complete the exercise

object varies. In general, the grasping time is longer as the object shape gets more complicated, e.g., grasping an apple is quicker than grabbing a felt-it pen. This trend can be attributed to the fact that interacting with an object with a simple shape requires less time since the exoskeleton better adapts to such forms. All the more so since the fingers are all moved simultaneously. It is also worth noting that lateral grasping arose to be more complicated than vertical one. This difference can be seen when comparing the water bottle with the tangerine or the tennis ball: the latter required indeed less time despite a similar diameter. The difficulties encountered by the subject to produce the correct sEMG pattern in the different positions might be attributed to the fact that the upper limb muscles differently compensate for the exoskeleton weight. It is hence evident how such cross-talk effects do modify the sEMG signal patterns.

Discussion—Beyond the chronometric data, these tests have also highlighted that executing a command action 50 Hz frequency results in a stressful operative condition both for the motor and for the user himself. The control system forced the motor to execute the commands coming from the classification loop every time they were different from the previous one. This condition has proven to occur quite frequently when the user (voluntarily or not) has found himself operating the exoskeleton with muscular signals mapped very close to the perimeter of the gestures polygons. In such

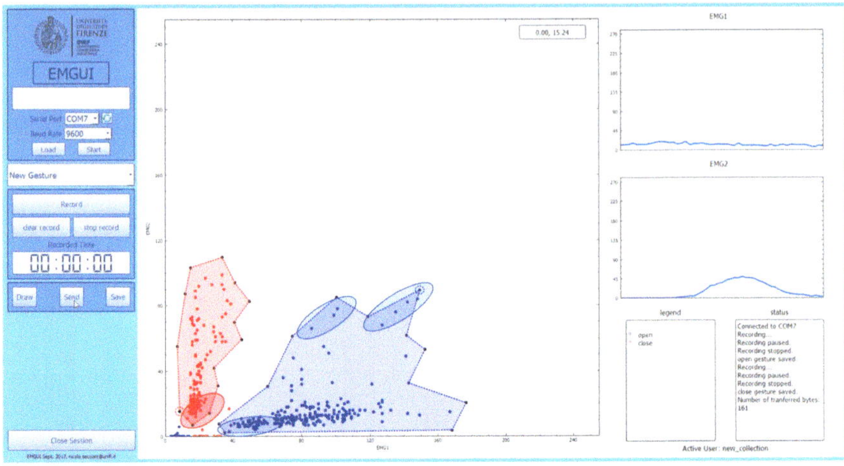

Fig. 4.10 The circled areas highlights some sample for which the control becomes glitchy. This happens because of the alternation between muscular states whose signals fall alternately in and out of the polygonal areas, producing a flickering transition from motion to stop

cases, the classified intention would likely jump from "closing" to "resting" and then back to "closing" very often, as visible in Fig. 4.10, even if the actual intention did not change. Hence, the motor was told to move (when the current test point was inside the polygon) and stop (when the test point was outside the polygon) up to 50 times per second. This working condition has ended up resulting in a glitchy actuation of the exoskeleton, which caused: (i) a quick overheating of the motor; (ii) a consequent loss of performance of the whole system; (iii) the onset of frustration in the user, fatigued by the vibrations and the problematic use.

These experiments have also confirmed that leaving the tuning of the classifier parameters to physiotherapists is efficient and suitable to the clinical application. During the training phase, it has been seen how choosing some key characteristics of the classifier—such as the size and position of the polygons—through direct interaction with the patient, and basing on his needs, produces positive results not only on the quality of control but also on the involvement of the patient in the exercise. Changing the size of the polygons affects the selectivity of the classifier: the larger the area underlying the polygons, the easier the user will be able to produce a muscular signal that falls within those figures. Conversely, this will increase the probability that such intention will be incorrectly classified when not desired. The sensitivity of the classification depends instead on the position of the polygons: the closer the polygon is to the origin, the less muscular effort is required to trigger the corresponding movement. However, it will also be more challenging to maintain muscle relaxation, such as not to incur erroneous classifications of movement intentions. Furthermore, by bringing both polygons closer to the origin and, therefore, even closer together, there is also a higher risk of particularly harmful incorrect classifications. E.g., classifying undesired open intentions when handling fragile objects or close intentions when touching a sharp or hot object. Therefore, these parameters must be chosen together with the patient so that he/she is not only more aware but also more in control of the strategy that will manage the HES. This process of mutual learning, by both the patient and the intelligent algorithm, is called "coadaptation" or "reciprocal adaptation" and is a highly studied subject in the field of human-robot interaction.

4.2 Second Test Session

The same experimental setup, experimental protocol, and classification strategy have been adopted during the second test phase. Instead, changes have been made to the actuation loop to mitigate the glitchy control issue arisen in the previous test session. The chain of instructions that translated the classified intention into the low-level control has been modified by adding a filtering action provided with a sort of memory of the last executed command and the last classified intentions. The classifier output was no longer translated directly into the corresponding motor command. Instead, it was added to a vector containing the last five classified user's intentions. Every time a new value entered the vector, the oldest one is removed, and all the others are moved in the following position. The vector size—five elements that correspond to a time window of 100 ms—has been chosen not to introduce too much latency and provide a still responsive control experience for a real-time application. A customized filtering action has been implemented to combine these five intentions and the last executed command—stored in a separate variable—to determine which action shall the HES perform; the main steps of its operation logic are detailed below.

Firstly, each classifiable intention has been assigned to an integer number (in), such as the rest state is the number in between the other two.

$$in(opening) = 1, \quad in(resting) = 2, \quad in(closing) = 3 \tag{4.1}$$

Secondly, a weight of 0.4 is assigned to the current executed intention, and other five weights—heuristically gathered from a unitary parabola for a total sum of 0.6—are assigned to the classified intentions. The smallest weight is matched to the oldest classified intention, while the most significant weights to the youngest one.

$$c_t \rightarrow w_c = 0.4 \tag{4.2}$$

$$i_{t-4} \rightarrow w_{t-4} = 0.011 \tag{4.3}$$

$$i_{t-3} \rightarrow w_{t-3} = 0.044 \tag{4.4}$$

$$i_{t-2} \rightarrow w_{t-2} = 0.098 \tag{4.5}$$

$$i_{t-1} \rightarrow w_{t-1} = 0.174 \tag{4.6}$$

$$i_t \rightarrow w_t = 0.273 \tag{4.7}$$

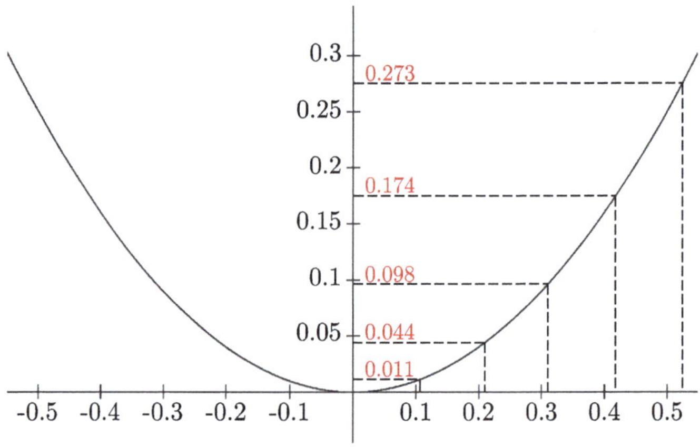

Fig. 4.11 The selection process of the five weights. These values—distributed along the y-axis and highlighted in red—are extracted from the equation of a unitary parabola $(y = x^2)$ so that their sum is equal to 0.6. Their corresponding abscissa are evenly spaced on the x-axis since the time interval between the classification of the different intentions remains constant

where c_t and w_c are the last executed intention at time t and its respective weight, i_{t-4}, ..., i_t the classified intentions at different time step t, and w_{t-4}, ..., w_t the corresponding weights, chosen as shown in Fig. 4.11.

Finally, a weighted average is calculated and rounded to the closest integer corresponding to an intention. The result of these calculations is then assumed as the new intention to be executed at time t (n_t).

$$n_t = w_c c_t + \sum_{k=0}^{4} w_{t-k} in_{t-k}(i_{t-k}) \tag{4.8}$$

where in_{t-k} are the numerical values assigned to the corresponding intentions i_{t-k}, accordingly to Eq. 4.1. The new intention is then assigned to the last executed intention variable for the following time step ($c_t = n_t$).

The results of this filter is a windowed smoothing action—window length = 100 ms, overlap = 20 ms—which has proved to reject single jumps from one intention to another. The filter is designed to give more importance to the latest classified intentions than to the oldest ones, and it provides the control system with a remarkable but not excessive inertia to changes: 40% of the decision on the new command depends in fact on the last intention performed by the exoskeleton ($w_c = 0.4$); the remaining 60% falls instead on the five intentions classified in the last time window of 100 ms and, in particular, roughly half of it depends exclusively on the newest one ($w_t = 0.273$). The selected weights are the result of laboratory tests, which highlighted how the modified control experience turns out to be much smoother, less power consuming, and less fatiguing for the user compared to the previous one. A comparison is shown in Fig. 4.12.

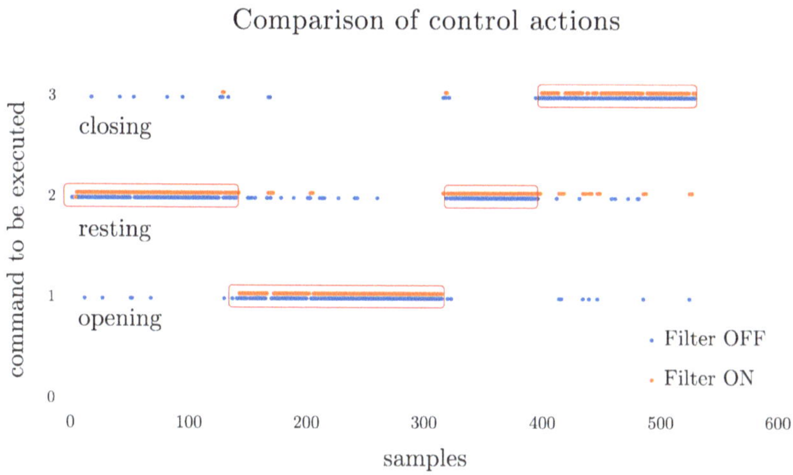

Fig. 4.12 The graph shows the difference between the resulting command to be executed generated by the old control strategy (in orange) and the one generated by the new control strategy after the implementation of the memory filtering (in blue) at a sampling frequency 50 Hz. The real actions to be reproduced are circumscribed in red. The intention labels (opening, closing, and resting) are assigned to y values according to Eq. 4.1

The figure shows a manifest decrease in the number of misclassified intentions, that is, all the points not inside the red rectangles. The most interesting result, though, is the excellent rejection of those mistakenly recognized classifications as the movement opposite to the real one (i.e., opening intentions classified as closing ones and vice versa). These are particularly harmful because, by imposing a sudden inversion of the exoskeleton motion, expose both the actuation system and the mechanical structure to significant stresses. Besides, as already mentioned, reproducing an intention that is opposite to the desired one, not only makes the control experience considerably annoying, but it can also be painful or dangerous for the user.

Results—The second test session has been carried out following the exact procedure of the first one and involving the same patient and physiotherapist: (i) the HES has been mounted on the subject's right upper limb by the physiotherapist; (ii) the subject has undergone the training phase for the classifier supervised by the physiotherapist; (iii) the control has been then tested on the virtual 3D hand model; (iv) the subject has finally carried out the pick-and-place chronometric task tests five times per object, and the average time to complete each task has been calculated. The results of this second test are shown below in Fig. 4.13, placed side by side with the previous test results.

Discussion—The evidence from the second test session shows that the introduction of the filtering action benefits the control experience and the effectiveness of the device. In almost all cases—except for the test that involved the tangerine, where the average time required to perform the task has increased—the results indicate

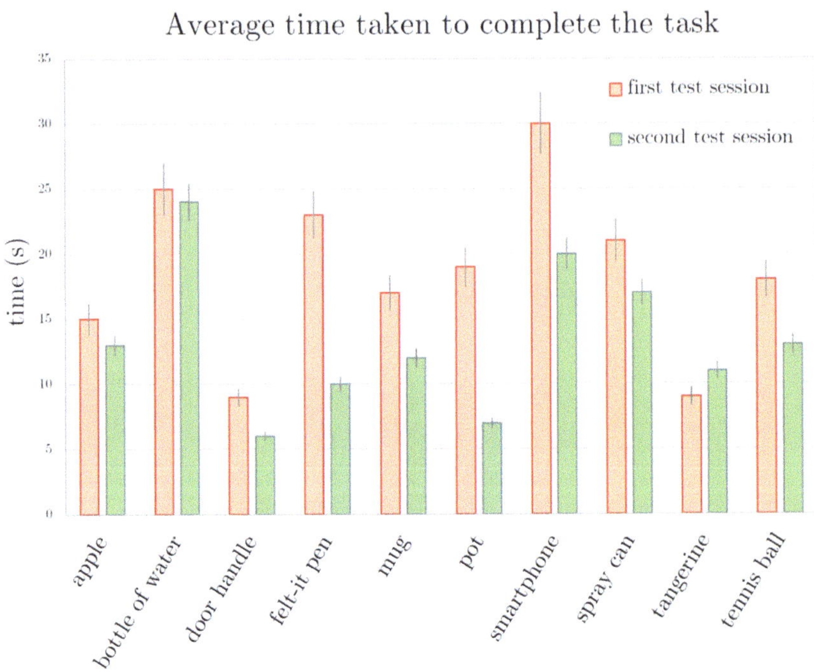

Fig. 4.13 The graph shows the results of the first and second test sessions, in orange and green respectively

that the new control strategy has positively impacted the time for completing the tasks. In some cases, such as the pot and felt-it pen, the time recorded has more than halved. It should be added that these improvements are not only due to the use of an improved control strategy. Although it has not yet been possible to carry out specific tests to verify the validity of such an assumption, it would, in fact, be unreasonable to exclude the sharing of other factors in improving the performance of the system. The factor that will undoubtedly have to be considered is the human factor. It is likely that, having already carried out a previous test session, both the subject and the physiotherapist have learned from the experience to better set up and use the system. Therefore, just as it is to be expected that the subject has become more familiar with the HES, it is fair to assume that also the physiotherapist has been able to take advantage of the previous experience to better tune the classifier on the subject's needs.

The last considerations concern the GUI developed for the tuning of the classifier parameters. The second test session has confirmed some of the strengths the graphical interface had already shown in the previous one. Firstly, its use proved to be particularly simple and intuitive: the four hours of teaching undergone by the physiotherapist who has supervised the tests have turned out to be more than sufficient to properly use the tool. Secondly, the time required for the system to be ready for the tests emerged to be remarkably short. In both sessions, it was never more than half an hour of classifier training. Finally—although the previous observation opened up scenarios where the classifier's training may take place before any daily use—it is possible to use the HES while keeping unchanged the parameters identified in a past training phase. This would require a training set sufficiently representative of the various use conditions produced by saving data from all the patient's training and loading them later in the same session to identify the polygons that fit the broad data set.

References

1. Secciani N, Bianchi M, Meschini A, Ridolfi A, Volpe Y, Governi L, Allotta B (2019) Assistive hand exoskeletons: the prototypes evolution at the university of florence. In: The international conference of IFToMM ITALY, pp 307–315
2. Secciani N, Bianchi M, Ridolfi A, Vannetti Yary Volpe F, Governi L, Bianchini M, Allotta B (2019) Tailor-made hand exoskeletons at the university of florence: from kinematics to mechatronic design. Machines 7(2):22
3. Bianchi M, Secciani N, Ridolfi A, Vannetti F, Pasquini G (2019) Kinematics-based strategy for the design of a pediatric hand exoskeleton prototype. In: The international conference of IFToMM ITALY, pp 501–508
4. Bianchi M, Secciani N, Ridolfi A, Vannetti F, Pasquini G (2019) Model-based approach in developing a hand exoskeleton for children: a preliminary study. In: International symposium on wearable robotics, pp 490–494
5. Matheus K, Dollar AM (2010) Benchmarking grasping and manipulation: properties of the objects of daily living. In: 2010 IEEE/RSJ international conference on intelligent robots and systems, pp 5020–5027
6. Charalambous CP (2014) Interrater reliability of a modified ashworth scale of muscle spasticity. In: Classic papers in orthopaedics, pp 415–417
7. Smith JD (2012) Single-case experimental designs: a systematic review of published research and current standards. Psychol Methods 17(4):510 December

Chapter 5
Conclusions

This work collects part of the research activities on wearable robotics at the Mechatronics and Dynamic Modelling Laboratory of the Department of Industrial Engineering of the University of Florence from 2016 to 2019. The content of these pages focused on developing a control strategy for an assistive hand exoskeleton system based on the exploitation of surface electromyographic signals. Paving the way to innovative low-cost solutions in robotic assistance for people with hand disabilities has constituted the main subject of this work. The motivation lies in the fact that people who have lost—wholly or partially—hand dexterity are deeply compromised in the possibility of living an independent and healthy life. Under the pressure of substantial technological progress in the middle of the Fourth Industrial Revolution, and with the World Health Organization indicating a clear positive trend in the number of people who will need assistance in the short term future, robotic devices nowadays represent a valuable tool to democratize the recovery of a better quality of life.

The overall research activity aimed to present a novel sEMG-based strategy to control a lightweight ($\simeq$500 g) and low cost ($\simeq$500 €) fully wearable assistive HES and describe the patient-centered process that has led to its implementation on an embedded system. The focal point was the choice of the classifier algorithm for the recognition of the user's motor intentions to be reproduced by the exoskeleton. The novelty of this work lies in the fact that, compared to related works available in the literature (Chap. 1) and previously implemented control strategy (Chap. 2), it shows particular aspects that raise the bar of the current state of the art of wearable assistive technologies. Unlike other studies, the complexity of the classifier has been brought from a one-dimensional to a two-dimensional level exploiting a particular algorithm whose use in the field of wearable robotics is not reported in any other works. This lightweight decision-making algorithm showed valid performance in comparison with standard machine learning techniques (Chap. 3) and arose to be more robust to external disturbances and more flexible to the classification of different intentions compared to the one-dimensional strategies. Still, it remained suitable for embedded real-time applications and this is another crucial point of the work. Two test sessions have been performed on a real patient in a clinical environment to validate

N. Secciani, *sEMG-based Control Strategy for a Hand Exoskeleton System*,
Springer Theses, https://doi.org/10.1007/978-3-030-90283-4_5

the discussed approach and qualitatively assess the usability of the proposed control strategy during object handling (Chap. 4). Promising results have been obtained, proving the goodness of the presented solution for wearable robotics applications. Nonetheless, there is still room for improvement.

Through this research activity, a more mature state of the device has been reached. It is now possible to request approval from an ethics committee for the actual clinical trial. Indeed, enlarging the test sample will allow for a proper statistical analysis for classification accuracy and object handling performance. It will also honestly assess if the same strategy can fit different users with different pathologies. In parallel, a variant of the actuation strategy will be tested to accelerate the grasping process and provide more natural hand motion control. The base idea will be to command the motor to move at speed proportional to the level of muscle contraction. Finally, the classification phase will be further enhanced. The development of a new classification procedure based on more sophisticated algorithms has already started to be designed. Microcontrollers technologies are evolving fast nowadays, and some manufacturers already provide tools for the implementation of pre-trained artificial neural networks directly in their microchips. The exponential growth of the computational power of the microcontrollers is opening, in fact, the doors for machine learning to wearable applications based on low-power embedded electronics.

All these topics, whose investigation constitutes a natural continuation of the presented research activity, will be subjected to further developments.

Author Biography

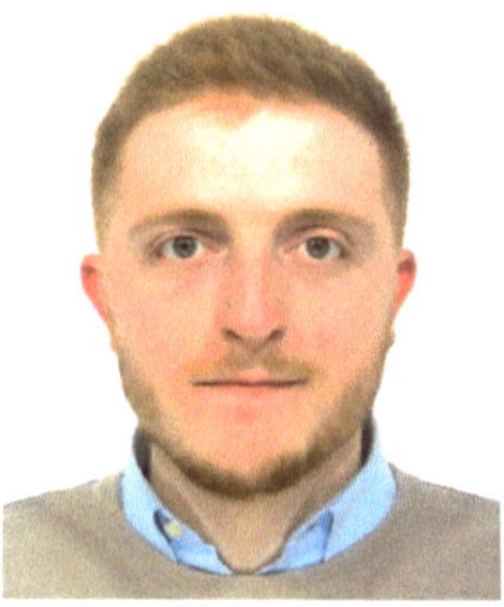

Dr. Nicola Secciani is currently a Postdoctoral Researcher at the Department of Industrial Engineering of the University of Florence, Italy (UNIFI DIEF), in Professor Benedetto Allotta's research group: the Mechatronics and Dynamic Modeling Laboratory (MDM Lab). Before his current role at UNIFI DIEF, Nicola received a Bachelor's degree in Mechanical Engineering in 2014 and a Master's degree in Electrical and Automation Engineering in 2016, both from the School of Engineering of the University of Florence, Italy. His Master's thesis has been carried out in collaboration with the Scottish Microelectronics Centre (SMC) at The University of Edinburgh, where Nicola spent four months as an intern. Then, he joined the MDM Lab for three years as a PhD student investigating original solutions to control wearable hand exoskeletons for assistive purposes.

This book is the result of Nicola's Ph.D. thesis. It collects the primary outcomes of his activity, which resulted in multiple publications in peer-reviewed international journals and conference papers, and in the awarding of his Doctoral degree in Industrial and Reliability Engineering in April 2020.

His current research interests focus on control strategies based on surface electromyographic signal classification for wearable robotic devices.

N. Secciani, *sEMG-based Control Strategy for a Hand Exoskeleton System*, Springer Theses, https://doi.org/10.1007/978-3-030-90283-4

MIX
Papier aus verantwortungsvollen Quellen
Paper from responsible sources
FSC® C105338

If you have any concerns about our products,
you can contact us on
ProductSafety@springernature.com

In case Publisher is established outside the EU,
the EU authorized representative is:
Springer Nature Customer Service Center GmbH
Europaplatz 3, 69115 Heidelberg, Germany

Printed by Libri Plureos GmbH
in Hamburg, Germany